NORTHERN SAN DIEGO COUNTY LAGOONS
Historical Ecology Investigation
REGIONAL PATTERNS, LOCAL DIVERSITY, AND LANDSCAPE TRAJECTORIES

PREPARED FOR THE STATE COASTAL CONSERVANCY

SEPTEMBER 2014

Prepared by:

Erin Beller[1]

Sean Baumgarten[1]

Robin Grossinger[1]

Travis Longcore[2]

Eric Stein[3]

Shawna Dark[4]

Scott Dusterhoff[1]

DESIGN AND PRODUCTION • RUTH ASKEVOLD[1]

[1] San Francisco Estuary Institute
[2] University of Southern California
[3] Southern California Coastal Water Research Project
[4] California State University at Northridge

San Francisco Estuary Institute Publication #722

SCCWRP Technical Report #831

Suggested citation:

Beller EE, Baumgarten SA, Grossinger RM, Longcore TR, Stein ED, Dark SJ, Dusterhoff SR. 2014. Northern San Diego County Lagoons Historical Ecology Investigation: Regional Patterns, Local Diversity, and Landscape Trajectories. Prepared for the State Coastal Conservancy. SFEI Publication #722, San Francisco Estuary Institute, Richmond, CA.

Report and GIS layers are available on SFEI's website, at http://www.sfei.org/ HE_San_Diego_Lagoons.

Front cover graphics: **Top**, Los Peñasquitos Lagoon framed by Torrey pines, ca. 1910 (John & Jane Adams Postcard Collection, courtesy of Special Collections & University Archives, San Diego State University Library). **Bottom,** Buena Vista Lagoon as shown on the T-sheet, 1887-8 (Rodgers and McGrath 1887-8b; courtesy of NOAA). **Back cover** (clockwise from top left): Buena Vista Lagoon, Agua Hedionda Lagoon, Batiquitos Lagoon, Los Peñasquitos Lagoon, San Dieguito Lagoon, and San Elijo Lagoon, January 2013 (photos by Sean Baumgarten).

TABLE OF CONTENTS

Acknowledgments vi

1. Project Summary 1

Project Background and Objectives 2

Study Area 6

Summary of Findings: Key Points 8

Summary of Findings: Historical Lagoon Ecological Mosaics 10

Summary of Findings: Contemporary Lagoon Ecological Mosaics 12

Summary of Findings: Variability in Lagoon Character 14

Application of Findings 16

2. Methodology 21

Data Collection and Compilation 21

Data Interpretation 22

Mapping Methodology 28

Mapping Classification 30

3. Regional Context..33

Environmental Setting 33

Land and Water Use History 38

4. Buena Vista Lagoon 50

Timeline 52

Deconstructing the T-sheet 54

Historical Synthesis Overview 56

Points of Interest 58

Salt Flat 60

Change Over Time 64

5. Agua Hedionda Lagoon 66

Timeline 69

Deconstructing the T-sheet 70

Historical Synthesis Overview 72

Points of Interest 74

Opening the Inlet 76

Change Over Time 78

6. Batiquitos Lagoon 80

Timeline 82

Deconstructing the T-sheet 84

Historical Synthesis Overview 86

Points of Interest 88

Habitat Type Configurations 90

Change Over Time 92

7. San Elijo Lagoon 94

Timeline 96

Deconstructing the T-sheet 98

Historical Synthesis Overview 100

Points of Interest 102

Salt Flat 104

Spatial Variability in Inlet Location 106

Salt Marsh and Salt Flat 108

Change Over Time 110

8. San Dieguito Lagoon 112

Timeline 114

Deconstructing the T-sheet 116

Historical Synthesis Overview 118

Points of Interest 120

Panorama of Salt Marsh and Channels..122

San Dieguito River and Valley 124

Change Over Time 126

9. Los Peñasquitos Lagoon 130

Timeline 133

Deconstructing the T-sheet 134

Historical Synthesis Overview 136

Points of Interest 138

Panorama of the Estuary 140

Postcards of the Lagoon 142

Change Over Time 144

10. Regional Synthesis: Ecological Patterns and Change 147

Regional Ecological Patterns...147

Ecological Functions 155

Habitat Type Change Analysis..162

Change Over Time 169

11. Regional Synthesis: Physical Patterns and Processes 171

Watershed Dynamics 171

Inlet Dynamics..176

Conceptual Synthesis 190

Recommended Future Research 200

References 201

BOXED TEXT

Why Historical Ecology? 4

Key Historical Data Sources for North County Lagoons 24

Edith Purer in Northern San Diego County 26

Lagoons, Sloughs, and Sea Swamps...27

From Glittering Salinas to Barren Salt Flats: Changing Perceptions of Salt Flats............................152

Nesting on the Salt Flat: California Least Terns and Western Snowy Plovers 160

Tidewater Goby 161

Water Quality Attributes 172

What Were North County Lagoons Like Before the Written Record? 174

Case Study: San Dieguito Lagoon Inlet Dynamics...182

Acknowledgments

This project was funded by the California State Coastal Conservancy. We would like to extend our deepest thanks to Megan Cooper, who helped develop the vision for the project and provided the support and resources that allowed for its successful completion.

The report benefited greatly from the input of our technical review team. Technical reviewers Roger Byrne, Wayne Engstrom, Letitia Grenier, Dave Jacobs, John Largier, Tony Orme, and Philip Williams each provided insightful guidance and comments on project mapping, analyses, and reporting. We would like to thank SFEI staff and colleagues, including Josh Collins, Linda Wanczyk, and Kristen Cayce, for assisting with analysis, interpretation, and graphical development. In addition, three interns from the Bill Lane Center for the American West at Stanford, Alexandra Peers, Rachel Powell, and Jenny Rempel, contributed to the project.

We are indebted to the staff and volunteers at the regional, state, and national archives and institutions that we visited over the course of the project, including the UC Berkeley Earth Sciences and Map Library, The Bancroft Library, The Huntington Library, Water Resources Collections and Archives, UCLA Benjamin and Gladys Thomas Air Photo Archives, National Archives and Records Administration Pacific Southwest Region, CSU Northridge Oviatt Library, Santa Barbara Mission Archive-Library, Seaver Center for Western History Research at the Los Angeles Museum of Natural History, Hearst Anthropology Museum, California Historical Society, Society of California Pioneers, California Language Archive, Bureau of Land Management, California State Railroad Museum Library, Stanford University Green and Branner libraries, and the Smithsonian Institution.

We also owe a great thanks to the staff and volunteers at numerous local historical societies and archives, including the San Diego History Center, San Diego State University Malcolm A. Love Library and Special Collections, UC San Diego Mandeville Department of Special Collections, UC San Diego Scripps Institution Archives, Oceanside Public Library, Oceanside Historical Society, Carlsbad City Library's Carlsbad History Room, Encinitas Historical Society, San Dieguito Heritage Museum, San Diego Public Library, San Diego County Cartographic Services, and San Diego County Assessor, Recorder, and County Clerk.

The dedicated staff at local lagoon foundations were critical to the success of this project. In particular we would like to thank Doug Gibson at the San Elijo Lagoon Conservancy and Mike Hastings at the Los Peñasquitos Lagoon Foundation for sharing their expertise on local ecology and history and giving us access to repositories of contemporary and archival data. We also received input from Ron Wootton at the Buena Vista Lagoon Foundation, Lisa Rodman at the Agua Hedionda Lagoon Foundation, Fred Sandquist at the Batiquitos Lagoon Foundation, and Shawna Anderson at the San Dieguito River Park.

Thanks to Keith Greer at SANDAG for his insights and support throughout the course of the project. Patricia Masters generously shared her expertise on southern California archaeology and paleo-coastlines. Thank you to Mark Miller at California State Parks and Mike Hastings at Los Peñasquitos Lagoon Foundation for taking us on a field tour of Los Peñasquitos Lagoon. Thanks also to Darren Smith at California State Parks for his helpful input.

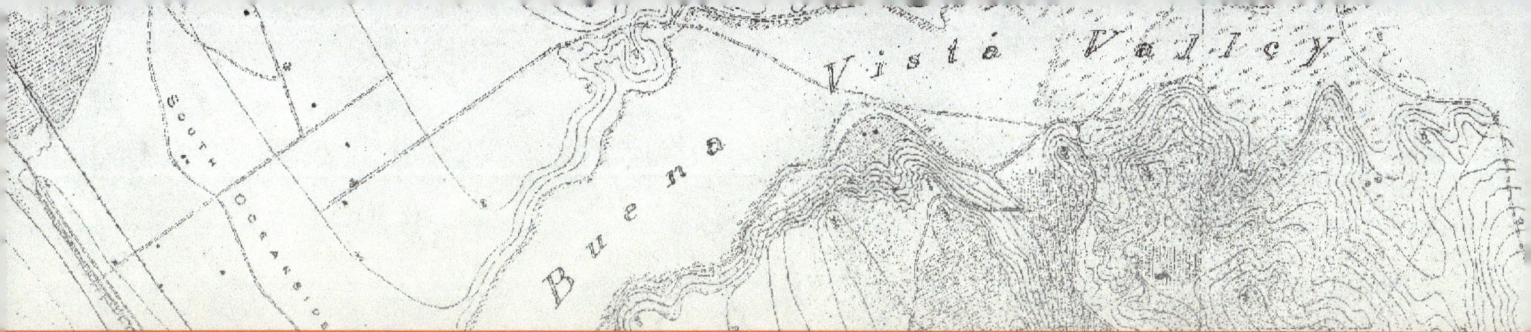

1. PROJECT SUMMARY

*D*riving between Oceanside and San Diego on Interstate 5, one can't help but notice the scenic expanses of water and marsh crossed by the freeway. These six estuaries – Buena Vista, Agua Hedionda, Batiquitos, San Elijo, San Dieguito, and Los Peñasquitos lagoons, each occupying a valley cut into the marine terraces of San Diego County – are an extremely important coastal wetland resource for the southern California region. They are valuable ecosystems both for native wildlife and for the people who live and recreate in and around their edges.

Compared to the extensive loss of coastal wetlands in neighboring areas, northern San Diego County ("North County") estuaries have remained remarkably protected over the past decades. This study finds that North County lagoons have lost only about 15% of their former estuarine area since the 19th century, a significant but relatively modest decline in the context of estimated regional losses of about half of total estuarine area across Southern California coastal systems (Stein et al. 2014).

At the same time, however, North County lagoons have experienced profound and widespread transformations as a result of impacts from a variety of land uses. Habitat loss and conversion have in many cases dramatically altered the ecosystem and social services provided by these estuaries. In addition, lagoon ecosystems have been degraded by an array of activities, including dredging and filling, the construction of transportation infrastructure, discharge of sewage effluent and other pollutants, dam construction and groundwater pumping, and urbanization. These and other anthropogenic modifications have heavily impacted the lagoons' character, including ecological patterns, water quality, tidal exchange, and freshwater inputs.

Today, these estuaries are the focus of numerous restoration and management efforts that aim to enhance lagoon function by reducing flooding, increasing tidal circulation, and increasing the acreage and quality of wildlife habitat, among many other objectives. As the region's scientists and managers take advantage of the significant opportunities presented by these systems, they face challenging decisions about what the goals of restoration should be. The study of the past can help inform these decisions by providing valuable knowledge about system characteristics under more natural conditions, as well as an understanding of how these characteristics have changed over time in response to human alterations to the landscape. Understanding the interaction between the ecological mosaic and underlying topographic, climatic, and hydrologic gradients, how these habitats supported native species, and

how elements of the landscape have persisted or changed is key to designing and managing locally appropriate future systems that are flexible, adaptive, and resilient to dynamic environmental conditions.

Though the study of these systems' past characteristics is a key component of determining appropriate restoration objectives, to date there has been no consensus about the natural structure and function of northern San Diego County lagoons as they existed in the recent past. While previous studies have addressed some aspects of the region's paleoecology (e.g., Cole and Wahl 2000, Scott et al. 2011) and historical ecology (e.g., Mudie et al. 1974 and 1976, Phillips et al. 1978, Hubbs et al. 2008, Grossinger et al. 2011), there has been no integrative and spatially explicit assessment of regional historical ecological and hydrogeomorphic patterns and processes. Further, the natural hydrology and ecology of estuaries in small southern California watersheds in general has not been well studied (Grewell et al. 2007).

Project Background and Objectives

The *Northern San Diego County Lagoons Historical Ecology Investigation*, funded by the California State Coastal Conservancy, seeks to address this regional data gap by reconstructing the landscape and ecosystem characteristics of northern San Diego County lagoons prior to the major modifications of the late 19th and 20th centuries. The research presented here analyzes historical landscape conditions for six northern San Diego County estuaries, supplying foundational information at both the regional and system scale about how these estuaries looked and functioned in the recent past as well as how they have changed over time. The ultimate goal of this study is to provide a new tool and framework that, in combination with contemporary research and future projections, will support and guide restoration design, planning, and management of coastal wetland systems in northern San Diego County.

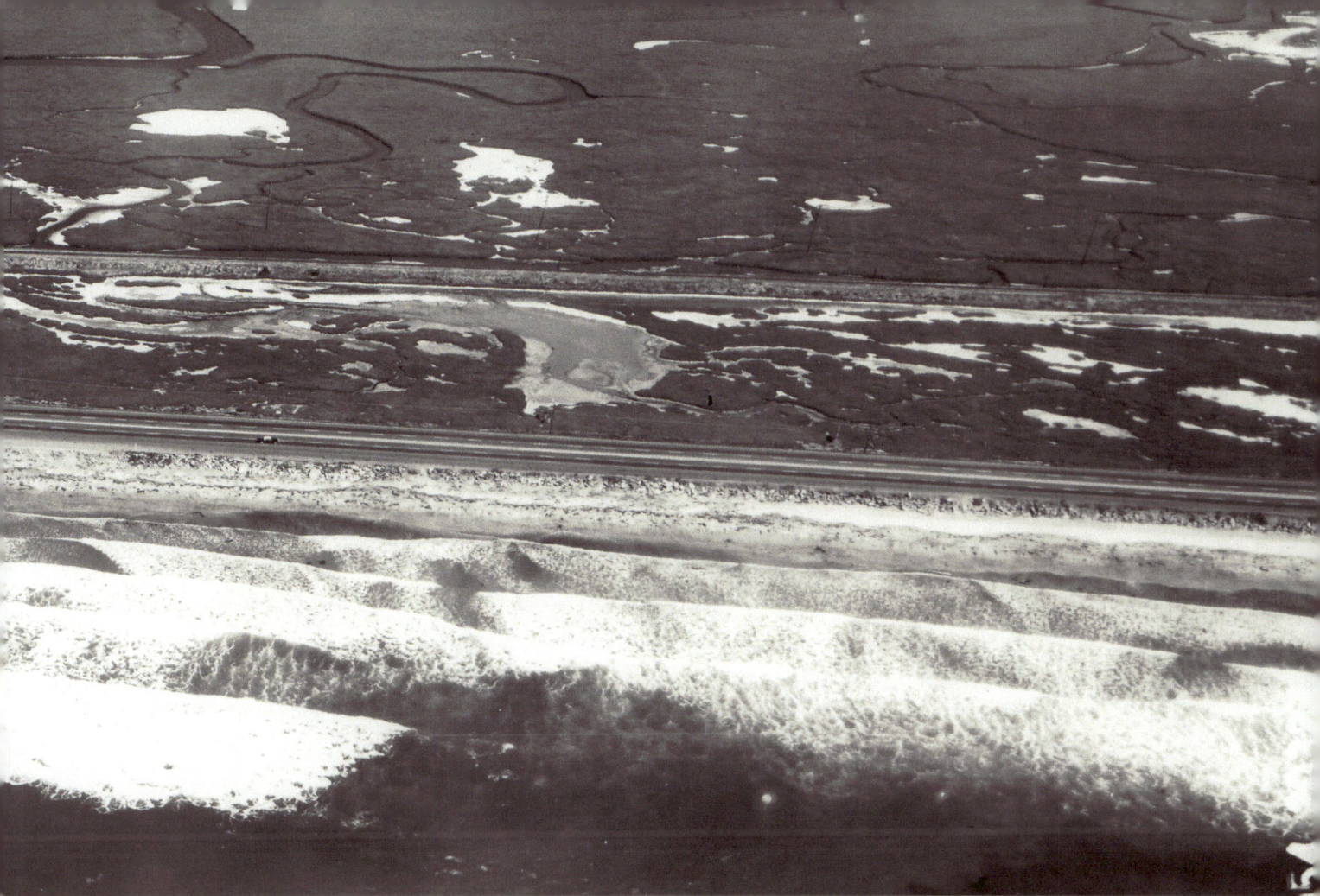

The study draws on a hundreds of historical documents to interpret and reconstruct the ecological and hydrogeomorphic characteristics of these estuaries circa the late 1700s to late 1800s, shortly after the arrival of Europeans (and thus the availability of written documents) but prior to subsequent large-scale landscape modifications. Data used in this report extend from 1769 through the 21st century, and range from travel diaries and family photographs to technical reports and government surveys. The resulting report and accompanying Geographic Information System (GIS) describe historical habitat type and distribution for each estuary, analyze hydrogeomorphic processes such as inlet dynamics, discuss driving physical processes, and quantify change over time.

San Elijo Lagoon, September 1954. (Collection 87-26, USA-C1 54 15/3, courtesy of Scripps Institution of Oceanography Archives, UC San Diego)

One of the primary products of this investigation is a map documenting historical habitat type patterns across the six North County estuaries (see pages 10-11). Information was compiled and synthesized in a GIS, which includes detailed map attributes such as historical sources and certainty levels for each feature. (See page 28 for mapping methodology; the geodatabase may be downloaded at www.sfei.org/he.)

This report complements the mapping with additional detail, context, and analysis. It is organized into eleven chapters: this chapter provides an overview of study goals and objectives and an introduction to the report. Chapter 2 (pages 21-32) provides a review of mapping and analytical methodology, and Chapter 3 (pages 33-49) summarizes the physical and historical environmental context for the region. Chapters 4 through 9 (pages 50-146) provide descriptions of the historical characteristics of each of the six estuaries. Chapter 10 (pages 147-170) integrates system-by-system findings into a summary of regional historical ecological patterns and change over time, while Chapter 11 (pages 171-200) explores key lagoon hydrogeomorphic characteristics such as inlet closure dynamics.

WHY HISTORICAL ECOLOGY?

The use of historical data to study past ecosystem characteristics is an interdisciplinary field referred to as "historical ecology" (Swetnam et al. 1999, Rhemtulla and Mladenoff 2007). Historical ecology is a powerful tool to reconstruct the form and function of past landscapes, enhancing our understanding of contemporary landscapes and helping us envision their future potential.

It can be tempting to see historical ecological research on one hand as an irrelevant exercise in nostalgia, or on the other as a restoration panacea, providing a prescriptive template from which to recreate the past. It is neither. Today's systems operate under different contexts than yesterday's, facing novel conditions from invasive species to climate change, and we could not to turn back the clock even if we wanted to. At the same time, many physical controls – from topography to geology – have remained relatively stable in many places, and history can provide relevant clues about how natural, resilient systems persisted in a particular place in the recent past. Historical ecology is not just about the "way things were," but also the way they worked, providing invaluable insight into system dynamics today (Safford et al. 2012a). A few points on the value of historical ecological research in supporting current planning and restoration efforts are briefly described below.

- Archival documents are a rich **dataset of locally relevant ecological information** that has been largely untapped in northern San Diego County, with the potential to change assumptions about past landscapes, document local and regional diversity, and link planning efforts to local heritage in a meaningful way.

- Historical ecology provides an opportunity to examine **system patterns, processes, and drivers at broad spatial and temporal scales**, describing the conditions to which native species are adapted and revealing fundamental characteristics and dynamics often difficult to discern in the contemporary landscape.

- Historical research can help foster a **shared understanding of local landscape history** and habitat values, establishing a common reference point across diverse stakeholders and contributing to a collective sense of place among the public. It can also serve as an effective educational and communication tool, and has been shown to make stakeholders more receptive to future changes in management (Hanley et al. 2009).

- Historical ecology is a critical component in identifying **locally appropriate restoration targets** (Jackson and Hobbs 2009). It provides the context needed to document change over time, using this understanding to recognize both the **constraints and opportunities** posed by the contemporary landscape. Even in places that have experienced substantial changes, history can help identify which elements of the system have persisted or changed over time, framing what may (or may not) be possible under new conditions (Higgs 2012). This can ultimately translate into project cost savings by revealing restoration strategies that are realistic for the site and would require minimal maintenance. Conversely, ignoring historical context can lead to inappropriate and ultimately unsuccessful restoration targets (e.g., Kondolf et al. 2001).

- Similarly, historical ecology can help us design and manage more flexible, **resilient future ecosystems** (Safford et al. 2012b). The study of historical landscapes can provide clues to how ecosystems were adapted to a highly variable, episodic climate regime, buffering the effects of environmental extremes while providing diverse ecological functions. As a result, historical ecology has particular relevance in the context of global climate change: as we anticipate a more variable future climate, we can learn from the ways in which intact dynamic ecosystems were able to respond and adapt to extreme, variable conditions in the recent past (Harris et al. 2006).

Of course, history is only one piece of the puzzle: restoration and management strategies must also incorporate a thorough understanding of contemporary conditions and future projections, as well as social and environmental values and objectives. However, historical ecology provides critical context and is an important consideration for any program that aims to restore biodiverse, resilient ecosystems.

A train crosses over Buena Vista Lagoon as a family looks on in this early 20th century photograph. Early landscape photos such as this one can provide important clues about the historical conditions of northern San Diego County's coastal estuaries. (photo #HP0673.001, courtesy of Carlsbad City Library Carlsbad History Room)

Study Area

This study examines six estuaries on the northern San Diego County coast: Buena Vista Lagoon, Agua Hedionda Lagoon, Batiquitos Lagoon, San Elijo Lagoon, San Dieguito Lagoon, and Los Peñasquitos Lagoon (collectively referred to here as "North County lagoons"). Each lagoon is situated at the mouth of a broad river valley cut into the surrounding marine terraces. The wetland complexes range in size from approximately 220 to 610 acres, and are separated from the ocean by barrier beaches that were historically breached with variable frequencies. The region is characterized by a Mediterranean climate with hot, dry summers and mild, wetter winters.

The study area extends approximately 18 miles along the coast from the northernmost lagoon (Buena Vista) to the southernmost (Los Peñasquitos), and includes portions of the cities of Oceanside, Carlsbad, Encinitas, Solana Beach, Del Mar, and San Diego/La Jolla. It encompasses the historical footprint of each lagoon complex, including both historical estuarine areas and transitional freshwater/brackish wetlands inland and immediately adjacent to the lagoons. Upland areas and tributary creeks are not included within the study area. We also excluded several larger river-mouth estuaries found to the north of Buena Vista Lagoon (notably, those of the San Luis Rey and Santa Margarita rivers) as well as a few smaller creek mouths in this region (e.g., Cottonwood Creek). The six systems studied were chosen both for their overall similarity (e.g., in watershed size, wave climate, and location), as well as to capture some of the region's estuarine diversity. Many are also the focus of extensive recent, ongoing, or planned management efforts.

People have lived along the San Diego County coast for at least 9,000 years. At the time of European contact in 1769, the San Diego County coastline was occupied by two tribes, the Kumeyaay (also referred to as Diegueño) and the Luiseño. From the late 18th through late 19th centuries, however, much of the northern San Diego County coastline was only sparsely populated. The relative aridity and isolation of North County meant that though many travelers passed through the area (often on their way between San Diego and Los Angeles, or Mission San Diego and Mission San Luis Rey), it experienced more limited settlement and agricultural activity compared to many neighboring coastal regions. Not until the construction of the California Southern Railroad in the late 1800s did the region begin to become accessible to larger numbers of people for the first time, eventually precipitating substantial changes in land use (see Chapter 3 for more information).

Today, more than 400,000 people reside in the six cities surrounding these lagoons (including La Jolla, but not all of San Diego), with many more thousands residing in their watersheds. The lagoons themselves are part of the urban fabric of North County, unmissable by anyone traveling across them via Highway 101 or Interstate 5. They currently supply a diverse array of social and environmental benefits: they are ecological and natural reserves; places for hiking, kayaking, and fishing; habitat for sensitive species, and sources of power plant cooling water – to name only a few services provided by different lagoons.

From Carlsbad south there are lagoons, sand dunes and broken ground until you reach the mouth of the San Marcos.

—LOS ANGELES HERALD 1887

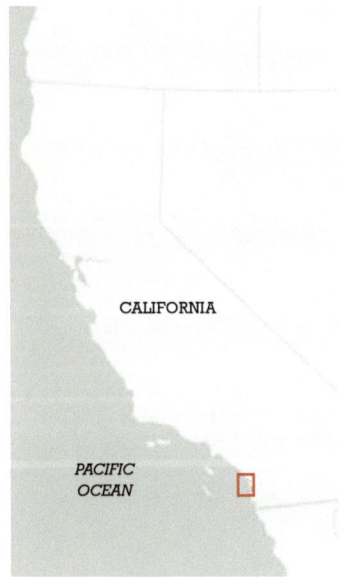

The project study area (shown in dark gray) encompasses six lagoons in northern San Diego County, and extends from Oceanside to Del Mar.

Summary of Findings: Key Points

The following pages summarize the key findings of the *Northern San Diego County Lagoons Historical Ecology Investigation*. These six systems – similar yet distinct, dynamic and resilient – were each adapted to the region's low-rainfall Mediterranean climate, hydrology, and sediment supply, and maintained their overall structure and function through the 18th and 19th centuries (and in some cases even much of 20th century) despite increasing land-use impacts. More detail on each of these themes can be found throughout the report.

- **North County lagoons supported a diverse array of habitat types in the recent past,** including salt marsh, seasonally flooded salt flat, mud flat, and open water in channels and ponds. While the relative proportions of each component varied from estuary to estuary, all systems were dominated by habitat types relatively high in the tidal frame (salt marsh and salt flat), which together constituted the great majority (95%, including ~50% salt marsh and ~45% salt flat) of the total estuarine area across the six systems. Open water and intertidal mud flat composed the remainder (~5%). Extensive freshwater/brackish wetland complexes were present at the back edge of each estuary, creating a gradual transition zone between estuarine and upland habitat types that in some cases extended several miles inland. These wetland complexes were composed of a matrix of seasonally and perennially flooded wetland habitat types reflecting a range of salinity tolerance.

- **Lagoons historically opened and closed with variable frequencies and to variable degrees.** While data are insufficient to quantify the precise closure regime of any system, it is clear that no lagoon was open all the time or closed all the time to tidal exchange in the recent past. Rather, each lagoon experienced a range of closure conditions, from open (including any state in which ocean water could enter the estuary, even if circulation was limited to a narrow tidal range) to closed, with inlet dynamics varying by season, year, and lagoon. This is in contrast to larger, fully tidal systems (such as San Diego Bay) as well as smaller systems that were rarely, if ever, tidal (such as the Ormond Beach wetlands; see Beller et al. 2011).

- **Highly dynamic, variable environmental conditions supported highly dynamic, variable habitat types.** North County estuaries were subject to changing environmental conditions on various time scales, including decadal climate cycles; interannual and seasonal variations in precipitation, flow, and inlet condition; and irregular seasonal and diurnal variability in tidal inundation. Exposure to these dynamic environmental conditions resulted in complex ecological patterns that varied across space and time, and were often very different than the more regular salinity gradients and plant zonation patterns observed in the few fully tidal estuaries in California (e.g., San Francisco Bay; Grewell et al. 2007). For example, a desiccated salt flat with pools of hypersaline water in August could be tranformed into a nearly freshwater lagoon a few months later, then a tidally flooded area once the inlet breached (see pages 14-15).

- **As a result, what we know as "lagoons" were not lagoons year-round.** These estuaries were not the deep, perennial water bodies conjured by the term "lagoon" – indeed, the term was not widely used in association with North County estuaries until the late 19th century. Instead, these systems were dominated by salt marsh and salt flats. The salt flats were only seasonally inundated, creating relatively shallow bodies of water during the wet season that dried out during the dry season, leaving large expanses of salt. Only in years where there was enough inflow for the salt flat to not dry out completely – yet not enough water to breach the inlet – did an open-water lagoon persist throughout the year. The little open water that did remain through the dry season often persisted in smaller, deeper ponds and channel segments within the marsh plain.

- **Despite this variability, many aspects of the lagoons persisted across decades and centuries.** Though these estuaries naturally exhibited dynamic conditions such as intermittent opening to tidal exchange and variable flooding frequencies, many fundamental landscape patterns remained relatively stable. For example, the presence of extensive salt flats was consistently documented across decades for many systems – in the case of Batiquitos Lagoon, for example, salt flat was documented to persist over a span of over 200 years.

- **The estuaries, though small, were part of a group of systems that contributed significantly to the richness and diversity of southern California coastal wetlands.** Buena Vista, Agua Hedionda, Batiquitos, and San Elijo each supported seasonally flooded salt flat on over 50% of their area, yet retained very limited perennial open water. Los Peñasquitos and San Dieguito Lagoons (along with Santa Margarita to the north) were salt marsh-dominant with limited intertidal flat and perennial open water areas. Many of these lagoons represent a relatively uncommon estuarine type not prevalent elsewhere historically in southern California: systems supporting significant areas of both salt marsh and salt flat, yet with relatively little perennial open water and intertidal flat area. Though they accounted for less than 10% of the estuarine area for the South Coast region, the supported approximately one-third of the region's salt flat habitat (Grossinger et al. 2011).

- **Within this group, no two lagoons were the same.** Though this population of systems experienced many of the same environmental conditions – similar climate, tidal and littoral regimes, and topography – and shared many of the same landscape-level characteristics, each estuary was different. They supported a range of habitat mosaics, from salt marsh-dominant with no salt flat (San Dieguito) to limited salt marsh and nearly 85% salt flat (Batiquitos). Historical evidence and tidal prism estimations suggest a range of closure conditions as well, with some systems with small watersheds (e.g., Buena Vista) likely more frequently closed to tidal circulation and others with larger watersheds (e.g., San Dieguito) likely open in at least a portion of the tidal range for a greater part of the year. This diversity represents only one segment of a much larger gradient of intermittently-closing systems (Jacobs et al. 2010).

- **Northern San Diego County lagoons have experienced significant transformations over the past centuries.** Changes in habitat type distribution have been driven by multiple factors, including dredging and inlet manipulation, changes in the timing and volume of freshwater and sediment inputs, and the fillilng and development of former wetlands. Seasonally flooded salt flats, which once covered over 1,200 acres across these six systems, now cover 90% less area than they did historically. They have been replaced – sometimes intentionally – by other habitat types, including subtidal open water, freshwater/brackish marsh, and salt marsh. Perennial open water, which historically comprised only a small (~5%) portion of these systems, has increased in extent by over 600%. Large areas historically occupied by transitional freshwater/brackish wetlands on the upland margins of the lagoons have decreased in extent by more than 50%, and in some instances have been almost completely lost to urban development.

- **Despite these changes, elements of the lagoon ecosystems have persisted to the present day.** The total area of salt marsh in the contemporary lagoons is only slightly less than the salt marsh extent historically, though there have been substantial shifts in the location of this habitat type within and among the lagoons. Many features within the marsh plains, such as ponds and channels, have also persisted, though they may be disconnected from the processes that formed and maintained them historically.

Summary of Findings: Historical Lagoon Ecological Mosaics
~late 1700s-late 1800s

Reconstruction of habitat types and ecological characteristics of six northern San Diego County lagoons, representing average dry-season conditions, prior to substantial Euro-American modification (~late 1700s-late 1800s). Salt marsh and seasonally flooded salt flat comprised the dominant estuarine habitat types across the six systems, while open water and intertidal mud flat occupied a more limited area. Freshwater/brackish transitional wetlands extended up the river valleys on the inland sides of the lagoons. More detailed views of each lagoon are shown in chapters four through nine.

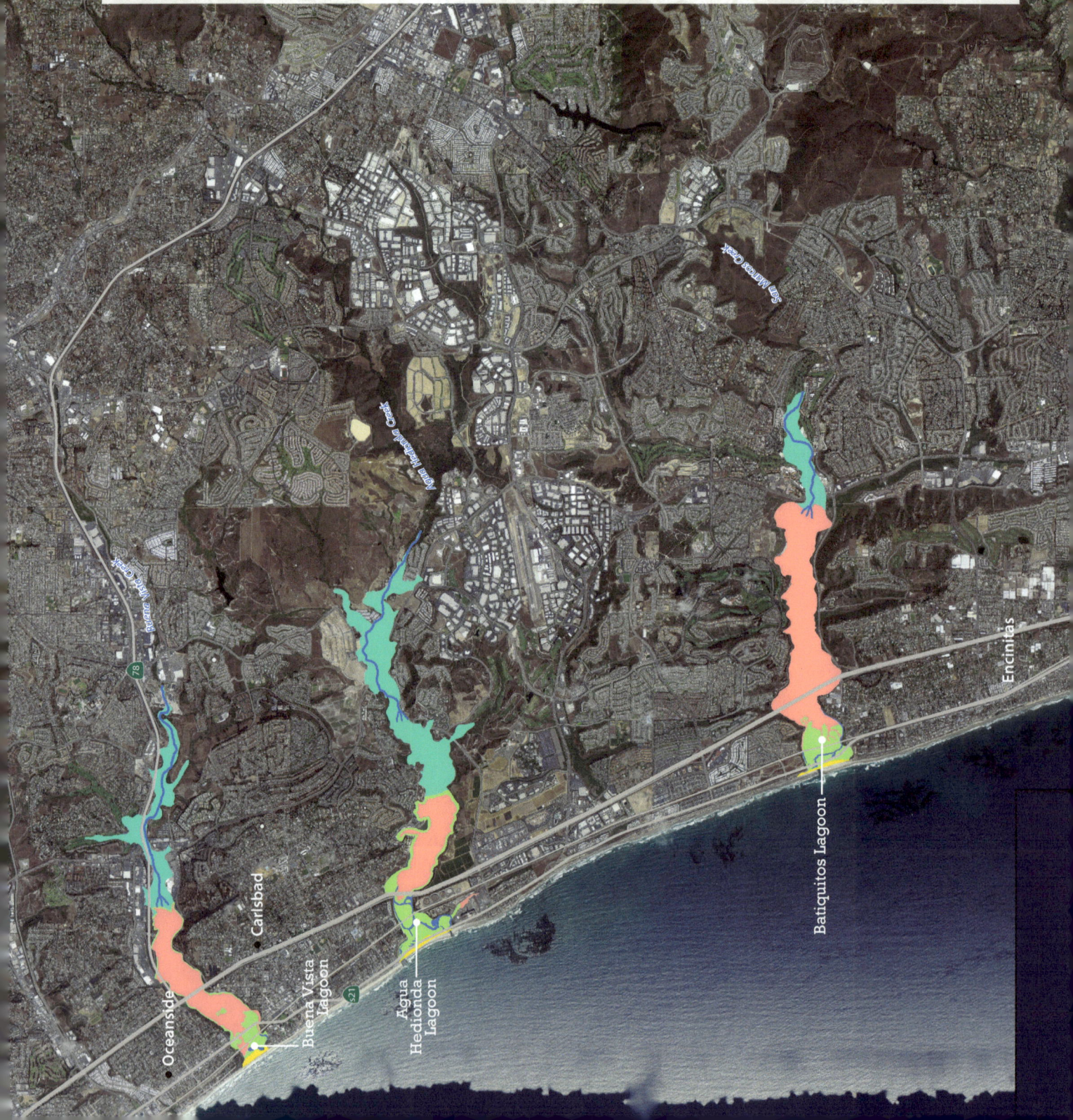

San Marcos Creek

Agua Hedionda Creek

Buena Vista Creek

78

21

Encinitas

Batiquitos Lagoon

Carlsbad

Buena Vista Lagoon

Agua Hedionda Lagoon

Oceanside

Rancho Santa Fe

San Dieguito River

Escondido Cr.

La Orilla Cr.

San Elijo Cr.

Carmel Creek

Los Peñasquitos Creek

56

5

Solana Beach

Del Mar

Cardiff By The Sea

S21

San Elijo Lagoon

San Dieguito Lagoon

Los Peñasquitos Lagoon

Freshwater / Brackish Wetland

Salt Marsh

Open Water / Mud Flat

Salt Flat (Seasonally Flooded)

Beach

Dune

Stream and Distributary

¼ mile

1:15,000

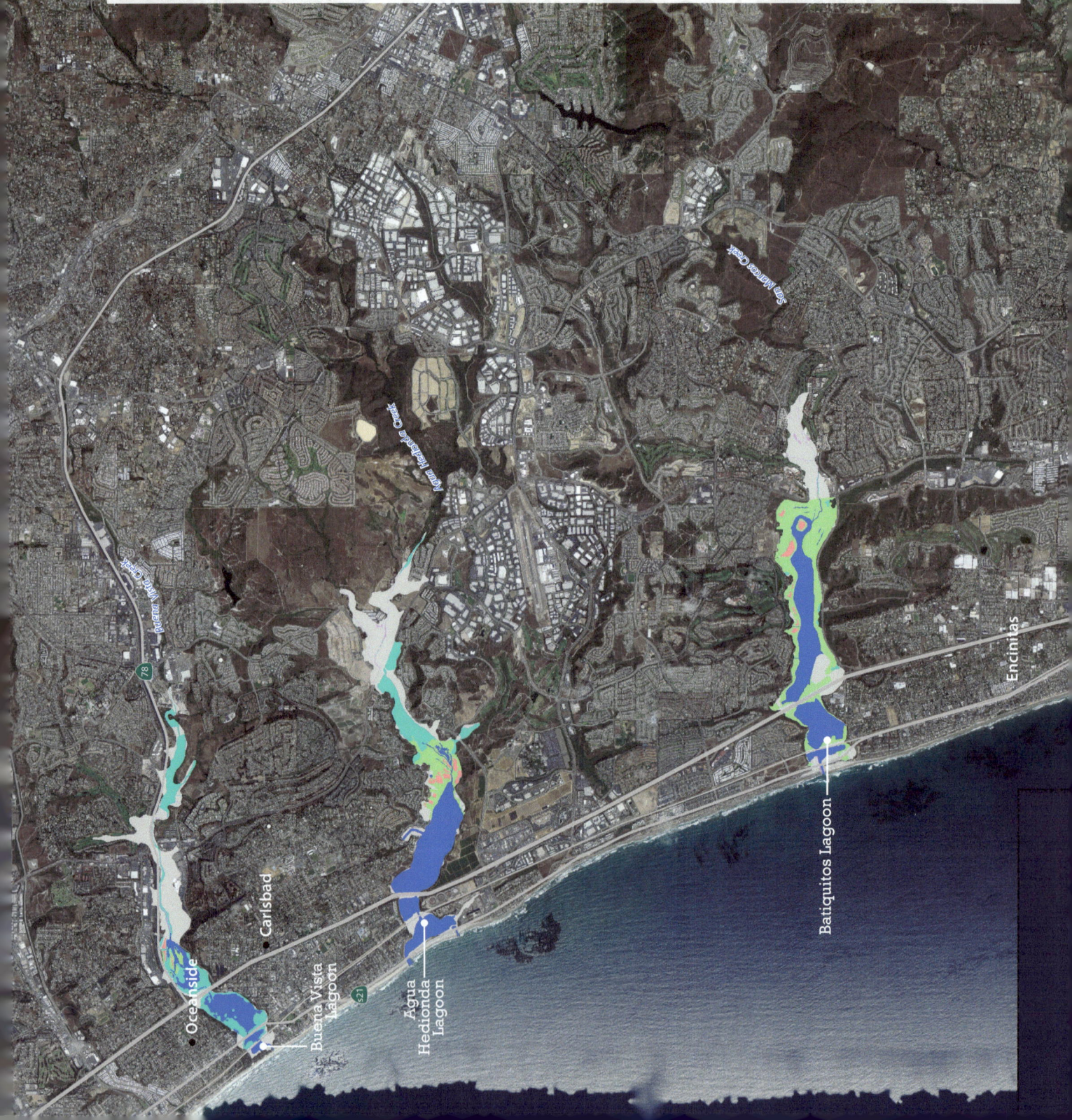

Summary of Findings: Contemporary Lagoon Ecological Mosaics
ca. 2010

Contemporary (ca. 2010) habitat mosaics characterizing northern San Diego County lagoons. Though some elements of the historical lagoon systems have persisted, many of the lagoons have experienced large-scale shifts in the distribution of wetland habitat types. Overall, the area occupied by salt flats and freshwater/brackish wetlands has decreased substantially, while perennial open water habitat has expanded. See page 162 for detailed information on mapping sources, methodology, and changes in habitat type distribution over time.

San Marcos Creek

Agua Hedionda Creek

Buena Vista Creek

78

Encinitas

Batiquitos Lagoon

Carlsbad

Oceanside

Buena Vista Lagoon

Agua Hedionda Lagoon

21

Rancho Santa Fe

Cardiff By The Sea

San Elijo Lagoon

Solana Beach

San Dieguito Lagoon

Del Mar

Los Peñasquitos Lagoon

Escondido Creek

La Orilla Creek

San Dieguito River

Carmel Creek

Los Peñasquitos Creek

Freshwater / Brackish Wetland

Open Water / Mud Flat

Salt Flat (Seasonally Flooded)

Salt Marsh

Developed

Other

¼ mile

1:15,000

N

Summary of Findings: Variability in Lagoon Character

Lagoon conditions varied inter- and intra-annually, tracking fluctuations in freshwater inflow, waves, and sediment delivery. This diagram depicts the cyclical variations in mouth state and flooding that characterized these system types. Though these variations tended to be seasonal, they would have been short-circuited by intra-seasonal fluctuations in stream flow or by unusually stable mouth states during anomalously wet or dry years, as depicted by dotted arrows in the center of the diagram. For example, in abnormally dry years a lagoon may not have filled sufficiently to breach, while in abnormally wet years it could have remained open for much of the dry season. (See pages 192-199 for more information on each phase.)

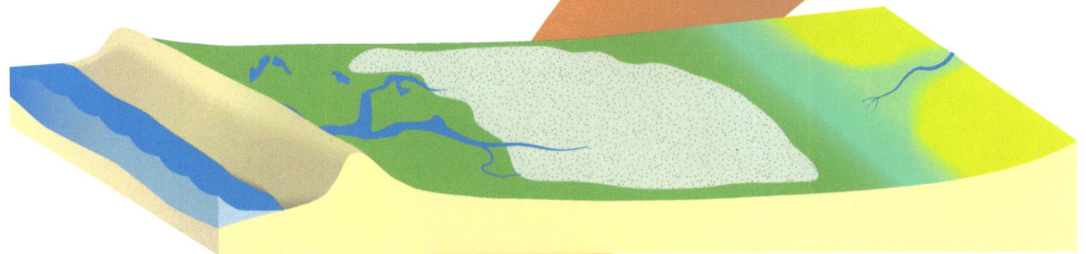

DRY PHASE (LATE SUMMER & FALL):
inlet closed, low inflow, lagoon dries up

During the dry season, when the inlet was closed, low freshwater inflow coupled with high evaporation rates led to net water loss, a drop in water levels, and the drying out of the lagoon, yielding hypersaline conditions and crystallizing salts.

little inflow; no breach

floods before fully desiccates

WET TO DRY PHASE (SPRING & EARLY SUMMER): inlet closes

As seasonal inflow declined, wave action could again close the inlet, cutting off tidal exchange to the lagoon. Lagoon water levels would rise or fall depending on net water balance (inflow vs. evaporation). During dry winters this phase may have occurred in winter months, and in some years, dry-to-wet and wet phases may have returned before the dry summer season.

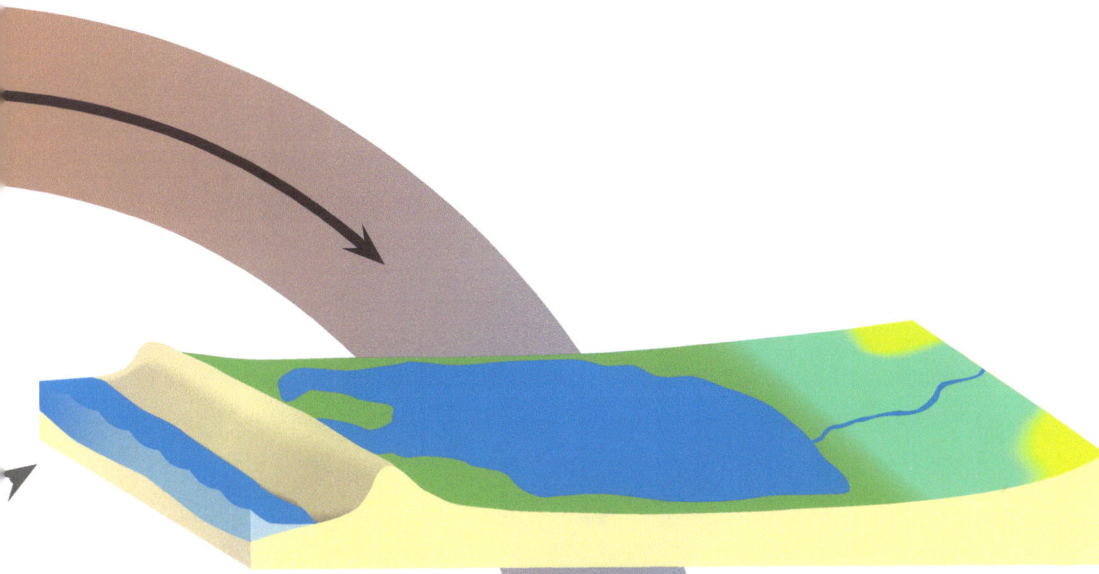

DRY TO WET PHASE (EARLY WINTER): inlet closed, stream flow fills lagoon

With the onset of rains, runoff would begin to fill the lagoon with fresh water, impounding behind the beach berm and often creating perched conditions (i.e., where the lagoon water level is above high tide). A freshwater/brackish lagoon would replace the hypersaline salt flat in the central portion of the estuary. Where more flooding space was available, the lagoon would persist longer.

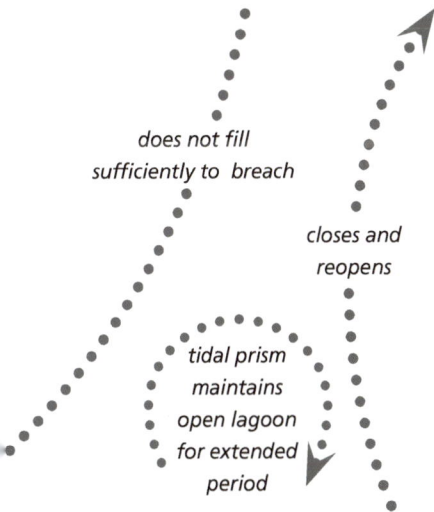

does not fill sufficiently to breach

closes and reopens

tidal prism maintains open lagoon for extended period

reopens

WET PHASE (MID- TO LATE WINTER): inlet opens, tidal conditions

Once sufficient water accumulated to overflow the beach berm, the beach barrier was breached and an inlet was formed, draining the lagoon and initiating tidal conditions. The beach barrier could also be breached by large waves overtopping the berm when lagoon water levels were very high. The lagoon would be subject to tidal exchange for a period of time. The duration of opening and the depth of the inlet channel varied, depending on the year (strength of inflow, occurrence of wave events) and system (available capacity to hold water, tidal prism volume, exposure to wave events).

Application of Findings

North County's lagoons have been many things to many people over the past centuries. They have been places to gather food and salt; they have provided swimming spots, race-tracks, and helicopter landing sites; and they have afforded striking views, parks, wildlife reserves, and trails, among a multitude of other benefits. These services shift as each generation brings a different vision of the lagoons' potential, based on the priorities, values, and context of their era. Some of these visions came to fruition and have shaped the lagoons as we know them today. Others remained unbuilt: for example, a 1928 proposal to turn one of the lagoons, which were determined to be "of very little value," into a freshwater lake for swimming, fishing, and boating (Oceanside Blade-Tribune 1928), or a 1947 proposal to turn the estuaries into marinas (see pages 18-19). What these systems are – and ideas of what they can and should be – has evolved over time, and will continue to do so.

The historical information presented in this report provides context and helps reveal challenges and opportunities in the contemporary landscape, but it alone cannot illuminate a path forward. Taken together with an understanding of current conditions and future projections, however, this information can provide perspective on the kinds of systems scientists and managers might seek to conserve and restore. The following points provide some considerations for managers, scientists, and the public as we imagine what the future might hold for these lagoons.

- **Recognize these dynamic, variable systems as natural estuarine types.** Intermittently closing, high intertidal elevation, salt flat and salt marsh dominant systems were functional, resilient estuaries that supported broad suites of native species. Their natural hydrology and ecology was different than that of California's few fully tidal estuaries, reflecting their specific hydrologic, geologic, and climatic context. The historical prevalence and ecological value of these system types has not been well recognized to date, particularly in the context of 20th century issues such as inlet constriction, decreased tidal prism, altered water quality, and increased sedimentation that can make system closure challenging and/or undesirable to manage.

- **Consider that native species were adapted to the dynamic patterns and processes of the past.** Lagoons provided a range of ecological functions that shifted by season and year, and species were adapted to cope with spatially heterogeneous and temporally fluctuating environmental conditions. For example, western snowy plovers, California least terns, tiger beetles, tidewater gobies, migratory waterfowl, and a multitude of other native wildlife used these lagoons for food and shelter. Habitat loss, conversion, and homogenization has greatly impacted the ecological functions provided by the lagoons (for more discussion see Ecological Functions section on page 155).

- **Incorporate remnants and analogs of formerly prevalent habitat types into restoration design.** Consideration of the types of habitats supported historically by these systems, evaluated within the context of courrent conditions and driving physical processes, will be instructive in determining what designs are likely to be resilient in the future. In addition, protecting remnants of former habitats where present may provide opportunities to support native wildlife and maintain ecological heterogeneity at a regional scale. Preservation and restoration of salt flats, transitional wetlands, and other habitat types, combined with management strategies that recognize North County's estuaries as dynamic systems that vary naturally over time, would enable these estuaries to better support native wildlife now and into the future by increasing biocomplexity and resilience.

- **Incorporate physical process into restoration design.** A detailed understanding of how former and current physical drivers have shaped, and continue to shape, these estuaries is just as important as understanding their former habitat patterns. In some places, this knowledge may allow for the identification and conservation of intact processes, or for the replication of these processes through restoration and management. Even in places where historical processes are no longer present, this understanding can be relevant to analog systems with comparable drivers, influencing how we manage them in the future.

- **Develop long-term restoration goals and strategies that account for likely future changes in environmental conditions.** Adaptation measures that anticipate climate and land use changes (e.g., sea level rise, altered precipitation patterns, and altered sediment and freshwater inputs) should be incorporated into restoration plans. Landscape patterns that conferred ecosystem resilience historically, such as physical gradients that can facilitate marsh migration, will be important for sustaining functional ecosystems in the future.

- **Manage for flexibility.** The lagoons were dynamic systems, subject to frequent disturbance and variable conditions on different time scales. Seasonal and interannual variability in inlet condition, inundation extent, and freshwater input were intrinsic to the systems. Reincorporating elements of this temporal variability and spatial heterogeneity in physical conditions and ecological patterns, rather than managing for stability, would make the systems more resilient to extreme events and anthropogenic changes and supply the range of ecological niches necessary for maintaining biocomplexity.

- **Think regionally, manage locally (promote diversity at many scales).** The systems studied here were part of a population of estuaries that were historically an important component of southern California estuarine diversity (Jacobs et al. 2010, Grossinger et al. 2011). This salt marsh/salt flat dominant estuarine type should be considered as part of a regional restoration palette. Each estuary was also unique, however, and different approaches should be considered for each system that take into account current differences in physical parameters. In other words, there is no "one size fits all" approach to restoration, even amongst these six estuaries.

(continued on page 20)

"Proposed Small Boat Harbors," 1947. Visions of the lagoons' futures have shifted over time. Here, a San Diego County Planning Commission map shows a proposed project – never built – to transform North County lagoons into marinas. A detail of the proposed plan for Agua Hedionda Lagoon is shown at right. (San Diego County Planning Commission 1947, courtesy of San Diego County Cartographic Services)

PROPOSED

SMALL BOAT HARBORS

SAN DIEGO COUNTY

COASTAL REGION

DEL MAR TO OCEANSIDE

San Diego County Planning Commission

1947

- **Reassess management goals that are based on sustaining fully tidal estuaries.** Some recent lagoon enhancement projects have focused on the excavation of subtidal habitat and the maximization of tidal prism. While many factors provide reasons for these activities – including infrastructure, habitat for marine life, and water quality issues – such features are generally not representative of historical lagoon conditions (ecology or hydrologic regime).

- **Integrate lagoon planning with watershed planning.** Many of the challenges associated with restoring flexible, dynamic estuaries can only be addressed by working at a watershed scale. Integrating upstream watershed management plans with downstream lagoon management will allow more flexibility in lagoon planning (e.g., with regard to water quality issues, restoration of transitional freshwater/brackish wetlands, or upslope migration with sea level rise).

2. METHODOLOGY

*A*rchival data offer a rich record of information about the historical ecologi-cal characteristics of northern San Diego County lagoons, and provide the raw data that form the foundation of this study. However, interpretation of these documents can be challenging given the broad range of authors, goals, techniques, and eras represented by the historical dataset (Harley 1989). Historical ecology methods must address the uncertainties associated with using this heterogeneous, non-standard array of data, integrating these diverse datasets into reliable and accurate landscape characterizations.

This chapter reviews the process through which historical data were discovered, interpreted, and synthesized for this study. For further detail on the methodology used to reconstruct historical landscape characteristics, please see Grossinger (2005), Grossinger et al. (2007), and Stein et al. (2010).

Data Collection and Compilation

Reconstructing historical landscapes requires a broad range of historical data, as a single dataset rarely provides sufficient information for accurate interpretation of complex systems (Grossinger and Askevold 2005). As a result, data collection constituted a significant component of project efforts. We visited 33 source institutions throughout California, including local and regional historical archives, county offices, and public and private libraries and museums (table 2.1, page 23). We also conducted searches of approximately 30 websites and electronic databases to obtain publicly available digital material. In total, we reviewed thousands of sources and collected a fraction of those reviewed.

Data collection efforts focused on the period from early Spanish explorers (1769) to the time of the first aerial photography in the late 1920s (see page 25). While this time period represents only a short time in the natural history of northern San Diego County lagoons, it is a relevant span for understanding how habitats were formed and maintained within a large-scale geomorphic and climatic context relatively similar to today's. This snapshot provides an opportunity not just reconstruct landscape patterns during the late 18th and 19th centuries, but also to understand the natural processes that shaped the distribution, diversity and abundance of habitats during this period – processes that in many cases may still be present.

Assembled data included maps (e.g., U.S. Geological Survey (USGS) topographic maps, U.S. Coast and Geodetic Survey (USCGS) T-sheets, Mexican land grant maps, county surveys, and soil surveys), photographs (plan view aerials, oblique aerials, and landscape photography), and textual documents (e.g., Spanish explorer accounts, travelogues and diary entries, General Land Office (GLO) surveys, and geology and water resources reports). We also drew from contemporary sources, including geologic maps, soil surveys, wetland and hydrology maps, elevation datasets, and modern aerial photography. While such datasets clearly depict an already-changed landscape, they can often reveal patterns that aid interpretation of the historical landscape when used in conjunction with earlier data.

Once collected, data were processed into more accessible formats for mapping and interpretation. A primary tool used in the data compilation process was a geographic information system (Esri's ArcGIS 10 software), which enabled the synthesis and comparison of many types of spatial data. We georeferenced a number of high-priority maps, including USGS topographic quads, railroad maps, GLO survey plats, soils maps, and county surveys. We also orthorectified and mosaicked the earliest available aerial imagery (about 200 images, taken in 1928-9) into a nearly continuous coverage of the study area. General Land Office survey data (over 1,400 data points) were also entered and digitized. Sources not compiled within the GIS (e.g., textual data, landscape and oblique photography, and maps too spatially imprecise to be georeferenced) were transcribed and/or organized by system and topic to allow for use of these data during interpretation and mapping.

Though the data collection process was extensive, it was inevitably not exhaustive. Undoubtedly, additional sources of information will surface in the future that will refine and enrich the understanding of landscape conditions presented in this report.

Data Interpretation

Constructing an accurate picture of historical landscape patterns and processes requires the integration, comparison, and interpretation of multiple independent sources (Grossinger and Askevold 2005). Sources were produced during different eras, using different methods and techniques, for differing purposes, and by different authors; consequently, an individual source considered alone may be ambiguous or misleading. The intercalibration of multiple data sources can uncover (and often resolve) inconsistencies between sources while at the same time revealing persistent features and patterns, ultimately ensuring more reliable mapping and interpretation. To take advantage of this synergy, we documented landscape features using multiple sources from varying years and authors wherever possible.

In particular, the use of relatively early (18th/19th century) and later (20th century) sources in combination often enabled the interpretation and mapping of historical features with a high level of accuracy and confidence, as detailed features visible in later sources often corresponded to features recorded in earlier, less accurate sources. For example, the 1920s aerial photographs show numerous channels, ponds, and other lagoon features which can also be discerned in 19th century maps. The early maps

Table 2.1. Source institutions from which data were collected for the northern San Diego County Lagoons historical ecology study. In addition to these archives, numerous online repositories were also consulted.

Source Institution	Location
Agua Hedionda Lagoon Foundation	Carlsbad
Buena Vista Lagoon Foundation	Carlsbad
Bureau of Land Management	Sacramento
California Historical Society	San Francisco
California Language Archive	Berkeley
California State Railroad Museum Library	Sacramento
Carlsbad City Library, Carlsbad History Room	Carlsbad
CSU Northridge Oviatt Library	Northridge
Encinitas Historical Society	Encinitas
Hearst Anthropology Museum	Berkeley
NARA's Pacific Southwest Region	Laguna Niguel/Riverside
Oceanside Historical Society	Oceanside
Oceanside Public Library	Oceanside
San Diego Archaeology Center	San Diego
San Diego County Assessor/Recorder/County Clerk	San Diego
San Diego County Cartographic Services	San Diego
San Diego History Center	San Diego
San Diego Public Library	San Diego
San Diego State University Malcolm A. Love Library and Special Collections	San Diego
San Dieguito Heritage Museum	Encinitas
San Elijo Lagoon Conservancy	Encinitas
Santa Barbara Mission Archive-Library	Santa Barbara
Seaver Center for Western History Research at the Los Angeles Museum of Natural History	Los Angeles
Smithsonian Institution	Washington, D.C.
Society of California Pioneers	San Francisco
Stanford University Library	Palo Alto
The Bancroft Library	Berkeley
The Huntington Library	San Marino
UC Berkeley Earth Sciences and Map Library	Berkeley
UC San Diego Mandeville Special Collections	San Diego
UCLA Benjamin and Gladys Thomas Air Photo Archives	Los Angeles
University of California San Diego Scripps Institution Archives	San Diego
Water Resources Collections and Archives	Riverside

KEY HISTORICAL DATA SOURCES FOR NORTH COUNTY LAGOONS

We drew upon a variety of historical cartographic, textual, and pictorial sources spanning many decades for mapping and interpretation. The summary below provides brief explanations of key datasets.

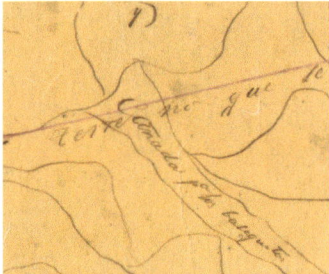

USDC ca. 1840a, courtesy of The Bancroft Library, UC Berkeley

U.S. Surveyor General's Office 1881, courtesy of Bureau of Land Management

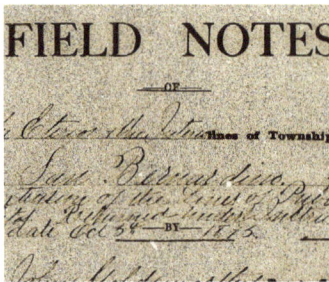

Goldsworthy 1874b, courtesy of Bureau of Land Management

Unknown 1888b, courtesy of California State Railroad Museum

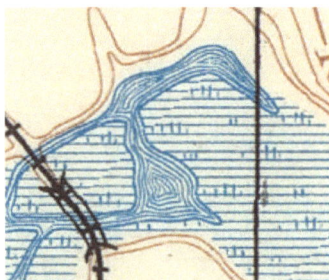

USGS [1891]1898

Mexican land grant sketches and court testimony (1840s-1860s). As the Mission system disintegrated in the 1830s, influential Mexican citizens submitted claims to the government for land grants. A *diseño*, or rough sketch of the solicited property, was included with each claim. Diseños often show notable physical landmarks which served as boundaries or natural resources, such as creeks, wetlands, springs, and forests. While diseños are not as spatially accurate as subsequent surveys, they provide extremely early glimpses of former landscape features and patterns.

General Land Office (GLO) Public Land surveys (1854-1914). In areas not claimed through the land grant system, the U.S. Public Land Survey imposed a grid of straight lines on the landscape, dividing property into six-mile square townships. Each township was further subdivided into 36 one-mile sections, each section containing 640 acres. Surveyors methodically surveyed section lines along these transects, noting cultural and natural features they encountered along the way. Survey notes and plat maps from these surveys are useful for their ecological information.

Textual accounts (1769-2013). Written accounts can provide a wealth of detailed information, with nuance about landscape dynamics not available on maps. Spanish expeditions provide the earliest accounts; later sources such as land grant case testimonies, newspaper articles, county histories, and travelogues give rich perspectives from early visitors and residents. Journal articles and technical reports from the 20th century were also mined for information.

California Southern Railroad maps (1881-1888). In 1881-82, a section of the California Southern Railroad was constructed along the San Diego coast from National City to Oceanside. Prior to and during railroad construction a series of survey maps were produced that depict many of the lagoons. Though the lagoon depictions are schematic in nature, they provide useful information about channel configurations and inlet conditions prior to the construction of the railroad.

U.S. Geological Survey topographic maps (1891-1980s). The USGS (established in 1879) began producing topographic quadrangles for North San Diego County in 1891. Though the maps are relatively coarse, they provide some of the earliest consistent, comprehensive coverage for the entire region.

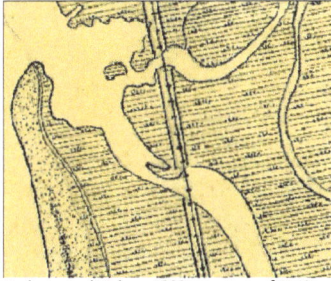
Rodgers and Nelson 1889, courtesy of National Oceanic and Atmospheric Administration (NOAA)

U.S. Coast and Geodetic Survey maps (1887-1934). The USCGS was established in 1807 to create navigation maps of the coastline and immediately adjacent areas. The maps covering the landward portion of the coastline, known as "topographic sheets" or "T-sheets," are a highly valuable source because of their large scale, remarkable detail, and high scientific standards. The three earliest T-sheets covering the study area were produced over several years by veteran surveyor Augustus F. Rodgers and two assistants, John Nelson and John E. McGrath. Rodgers and his colleagues spent several months mapping the region over a three-year period of generally average rainfall: August-December 1887 and May-November 1888 for the northern four lagoons, and May-July 1889 for San Dieguito and Los Peñasquitos. Though the T-sheets are invaluable, they are not without limitations. North County T-sheets were produced several decades later than T-sheets for other parts of the state, and therefore post-date the construction of the railroad by a few years. In addition, Rodgers' maps tend to show less detail (e.g., first-order channels) than maps produced by other USCGS surveyors (Grossinger 1995). For examples of annotated T-sheets and discussion of T-sheet symbols, see Grossinger et al. 2011 and Shalowitz 1964.

San Diego County 1928, courtesy of County of San Diego, Department of Public Works

Historical aerial photography (1928-29). The historical aerial imagery used in this study was taken during the winter of 1928-9, and represents the earliest complete coverage of the study area. While the photographs capture lagoon conditions after substantial modifications had already been made, they nevertheless reveal many relict ecological features and are extremely valuable for interpreting features depicted on earlier sources.

Storie and Carpenter 1929a

U.S. Department of Agriculture soil surveys (1915-29). Early soil surveys were developed to describe variability in the agricultural viability of regional soils. These maps, and their accompanying reports, are a key source in the inference of historical habitat extent and location. Descriptions of soil properties and agricultural use can provide insight into former habitats, in particular providing spatially accurate detail on wetland extent.

Harrington 1925, courtesy of National Anthropological Archives, Smithsonian Institution

Landscape photography (ca. 1895-present). Historical photographs represent a category of diverse historical data that can provide extremely localized, accurate information. Photographs can capture the conditions of a given place and time in a manner that provides substantial detail about specific species presence and landscape structure.

confirmed the historical presence of the features, while the aerial photos (or another spatially accurate source) allowed us to map them with a higher level of detail than would be possible using the early sources alone.

Data must also be interpreted within the context of climate and land use change. Knowing the season in which a particular source originated, or whether it was created during a wet or dry year, influences the interpretation of that source, and thus affects overall understanding of what constitutes "average" historical conditions. For example, an account of the San Dieguito River "[rushing] into the sea with great force" is more notable for having occurred in June of a dry year than it would have been had it been noted in January or in a year with above-average rainfall instead (Duhaut-Cilly [1827]1997). Likewise, the potential impacts of land use changes must be taken into account when evaluating a data source (see Chapter 3).

EDITH PURER IN NORTHERN SAN DIEGO COUNTY

In late autumn the marshes are beautiful with the red coloring from the anthocyanin in the succulent leaves of *Suaeda* and the stems of *Salicornia*.

—PURER 1942

DR. EDITH A. PURER
Senior Science
Botany

One of the most valuable papers from the 20th century technical literature is also one of the earliest: Dr. Edith Purer's "Plant Ecology of the Coastal Salt Marshlands of San Diego County, California". Published in *Ecological Monographs* in 1942, the paper describes the ecology and physical environment of twelve estuaries in San Diego County, from the Santa Margarita to the Tijuana River. It was a landmark paper not just for its descriptions of the region's salt marshes (to our knowledge, the first academic paper to treat the subject) and its pioneering research on factors influencing plant distribution and zonation in salt marshes (Zedler 2012), but also because of its author. Dr. Purer (1895-1990) earned her PhD from the University of Southern California in 1933 and was one of California's first female professional ecologists (Van de Hoek 2006). In addition to her dissertation on dune plants in southern California, Dr. Purer published eight peer-reviewed articles on the plant ecology of diverse habitats ranging from chaparral to vernal pools and coastal salt marshes. She was also an artist, and produced a number of paintings of landscapes around her San Diego home. Her research on the lagoons of northern San Diego County represents nearly three years of fieldwork beginning in 1938, and provides a uniquely early and rigorous glimpse into the character of these systems in the early 20th century.

LAGOONS, SLOUGHS, AND SEA SWAMPS

A recurring challenge of this study, and of historical ecology research in general, lies in translating the terminology used by different sources to describe features of the landscape into a contemporary classification framework. While small variations in word choice sometimes have little significance, in other cases subtle differences in language may reflect real physical distinctions or changes in the landscape. Furthermore, the meaning of a certain term may have evolved over time, such that uncritically applying the contemporary definition to historical sources may lead to erroneous interpretations. It is therefore important to try to understand the precise meaning of a term as it was used historically.

The usage history of the terms "lagoon" and "slough" provides an instructive example. The word "lagoon" was not widely used in association with the estuaries of San Diego County until the late 19th century. Instead, the lagoons were often referred to as "sloughs." In 1869, for example, GLO surveyor Pascoe describes coming upon the "edge of a large salt slough" and meandering "around head of same slough" at the eastern edge of Buena Vista Lagoon (Pascoe 1869). His location, along with the illustration in the accompanying GLO survey plat, make it clear that he is referring to the entire lagoon rather than an individual channel. In other instances, however, the usage of slough by early surveyors appears to be consistent with the contemporary definition of a channel within a wetland. For example, GLO surveyor Wheeler (1874-5) refers to a "slough 15 links [~10 feet] wide" on the edge of the Batiquitos Lagoon salt flat. There is some indication that the term "slough" was used by at least some 19th century observers to describe estuaries with "no large river emptying into it to keep it open" to the ocean (Alexander 1870 in Engstrom 2006).

The USCGS T-sheets (see page 25) do not use either term, instead simply labeling the areas around the estuaries as "valleys." Other terms used to refer to the lagoons in the past include "lagunas" (e.g., Wheeler et al. 1872), "esteros" (1769; Crespí and Brown 2001), "sea swamps" (Holder 1906), "swamps" (e.g., Hanson 1880), "flats" (e.g., Knox 1934b), and "natural salt lagoons" (Smythe 1908).

The challenge of deciphering the meaning of terms from historical sources is compounded by the fact that many of the first European travelers to California wrote in Spanish. Translations of the original Spanish are sometimes conflicting or misleading, and subtle differences in the translation can lead to significant differences in interpretation. An example is the translation of Friar Juan Crespí's use of the Spanish word *estero* in his journal during the Portolá expedition. As the expedition headed north from the San Dieguito River on July 15, 1769, Crespí described "*un estero bien grande de la mar*" (Crespí and Brown 2001). A leading translation of Crespí's journals translates *estero* as "inlet," but the term can also refer to an estuary or lagoon, which may be more appropriate in this context (Gudde and Bright 1998, Crespí and Brown 2001).

Mapping Methodology

Once collected, organized, and georeferenced, historical data were synthesized to create a map of historical landscape characteristics of the estuaries prior to major Euro-American modification. Rather than portraying the lagoons at a single point in time, the mapping is intended to represent average dry-season ecological conditions during the target period (~late 1700s-late 1800s). This reconstruction of the distribution of historical habitat types and channels is designed to serve as a tool for landscape interpretation, enhancing our understanding of regional ecological patterns and the processes that shaped them.

We used a GIS to integrate relevant data layers representing many disparate sources and time periods into a single layer, and to record attributes about each landscape feature (see facing page). To document the basis for the mapping and interpretation of each feature in the GIS, we attributed each feature with digitizing sources (i.e., sources used to digitize the feature) as well as any supporting interpretation sources (i.e., sources that verified or enhanced the interpretation of a feature). We did not attempt to document every piece of evidence that showed a given feature, but rather those that contributed most to its delineation and interpretation. Each feature was also assigned estimated certainty levels to indicate our confidence in that feature's historical presence and classification (interpretation), size, and location following standards discussed in Grossinger et al. (2007; table 2.2 below). Certainty levels were determined based on a combination of source date, accuracy of the digitizing source, diversity and quality of supporting evidence, and stability of features on a decadal scale. The application of attributes on a feature-by-feature basis allows users to assess the accuracy of different map elements and identify the original data, serving as a catalog of information sources (Grossinger 2005; see also Stein et al. 2010).

Table 2.2. Certainty levels. Each mapped feature was assigned a certainty level of high, medium, or low for each of three characteristics, following standards described in Grossinger et al. (2007). Interpretation describes our certainty that the habitat type assigned to the feature is accurate and that the feature existed historically. Size describes our certainty that the feature's spatial extent is accurately depicted. Location is our certainty that it existed in exactly that spot. Together these certainty levels help us record the uncertainties inherent in the mapping process.

Certainty Level	Interpretation	Size	Location
High/ "Definite"	Feature definitely present before Euro-American modification	Mapped feature expected to be 90%-110% of actual feature size	Expected maximum horizontal displacement less than 50 meters (150 ft)
Medium/ "Probable"	Feature probably present before Euro-American modification	Mapped feature expected to be 50%-200% of actual feature size	Expected maximum horizontal displacement less than 150 meters (500 ft)
Low/ "Possible"	Feature possibly present before Euro-American modification	Mapped feature expected to be 25%-400% of actual feature size	Expected maximum horizontal displacement less than 500 meters (1,600 ft)

We also developed a contemporary wetland map in order to analyze changes in habitat extent and distribution over time. The contermporary wetland map was compiled from regional wetland mapping developed by the Southern California Wetlands Mapping Project, additional local vegetation mapping (Greer and Stow 2003, Everest International Consultants, Inc. 2004, AECOM 2012), and modern aerial imagery (NAIP 2009). See page 162 for further discussion of the methodology used in the habitat change analysis.

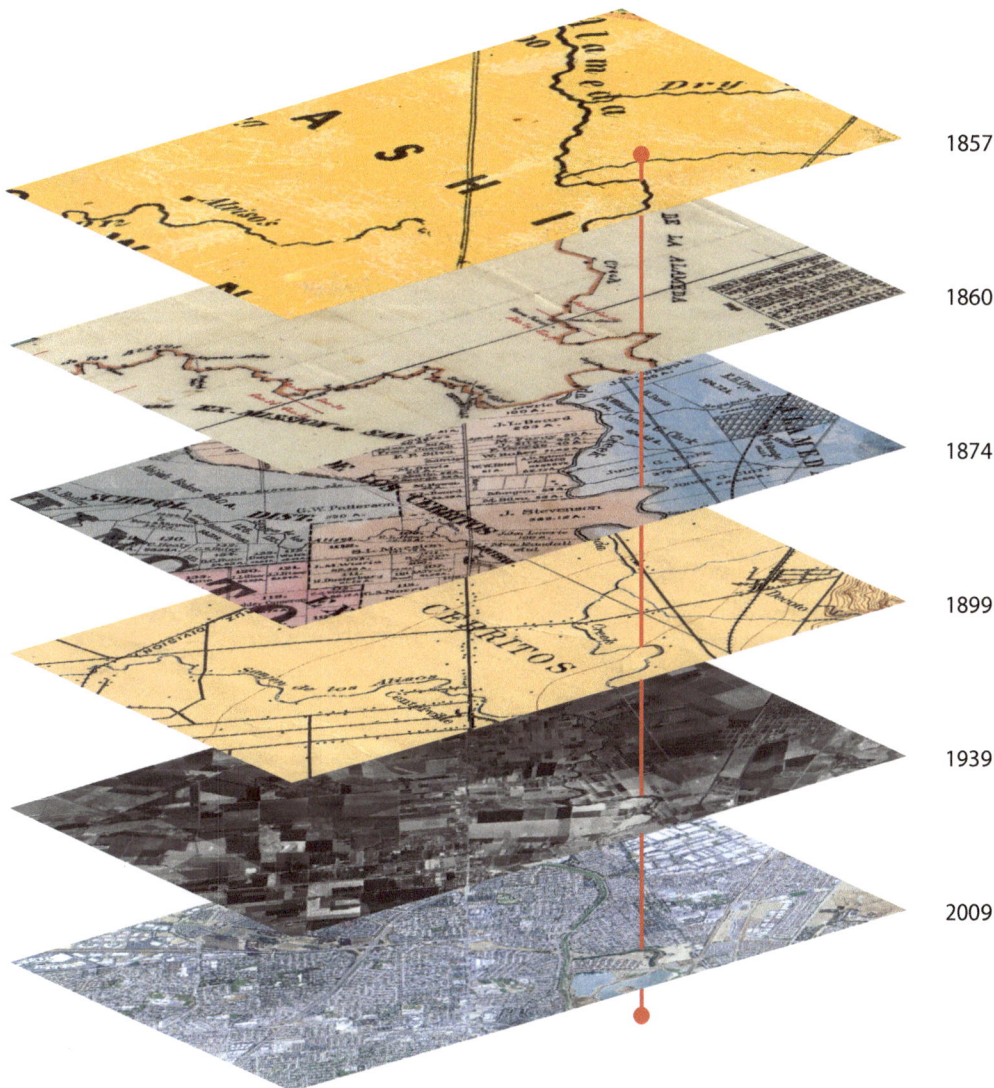

1857

1860

1874

1899

1939

2009

Maps assembled from different time periods shown in a geographic information system allows for comparison of features across space and time.

Mapping Classification

Our mapping utilizes six habitat types: beach, dune, salt marsh, open water/mud flat, salt flat (seasonally flooded), and freshwater/brackish wetland. We consider three of these classes (salt marsh, open water/mud flat, and salt flat (seasonally flooded)) to be "estuarine" habitat types. The classes are intended to capture broad-scale patterns and to be comparable with contemporary classification systems. Though they represent the greatest level of detail that could be mapped consistently across the study area from the available data, within each class there would have been complex fine-scale patterns and considerable variation in species assemblages. Two of the classes – open water/mud flat and salt flat (seasonally flooded) – reflect the inherent diurnal and seasonal variability of the lagoons. Brief definitions of each of the six classes are provided below (beach and dune are described together). For additional information about the salt marsh, salt flat, and freshwater/brackish wetland habitat types, see pages 149-154.

Beach/Dune. Beaches and dunes are coastal habitats located immediately along the shoreline. Beaches and foredunes are sandy and sparsely vegetated, while backdunes are located inland from foredunes and are generally more stable and more densely vegetated.

Salt Marsh. Salt marshes are wetlands dominated by salt-tolerant vegetation. The frequency of tidal inundation varies widely depending on elevation, inlet closure dynamics, and climate, causing spatial heterogeneity as well as wide temporal fluctuations in salt marsh salinity. Plant species distribution within salt marshes is determined by a variety of factors, including salinity, elevation, proximity to channels, and interspecific competition (Pennings and Callaway 1992, Zedler et al. 1999, Pennings et al. 2005). Common salt marsh plant species in north San Diego County's coastal lagoons likely included pickleweed (*Sarcocornia pacifica*), saltgrass (*Distichlis spicata*), shoregrass (*Monanthochloe littoralis*), Parish's glasswort (*Arthrocnemum subterminale*), and alkali heath (*Frankenia salina*) (Purer 1942, Grewell et al. 2007). This habitat type includes both large expanses of salt marsh as well as narrow fringes of salt-influenced vegetation that grew along the edges of salt flats, though these areas would have differed in terms of species composition.

Open Water/Mud Flat. Open water/mud flat includes unvegetated channels and ponds that are either permanently flooded or intermittently inundated by tidal fluctuations (i.e., both subtidal and intertidal areas). Because many of the lagoons were often closed to tidal influence historically, the inundation frequency of open water/mud flat did not necessarily follow daily tidal cycles.

Salt Flat (Seasonally Flooded). Salt flats are unvegetated or sparsely vegetated areas where high soil salinities generally preclude the growth of vegetation (Pennings and Bertness 1999). In San Diego County's coastal lagoons, salt flats trapped water during the rainy season, transforming them into expanses of open water. During the dry season, high evaporation rates caused the flats to once again dry out, concentrating salts. In some places, areas mapped as salt flats would have supported small patches of marsh vegetation that are not represented in the historical synthesis mapping.

Freshwater/Brackish Wetland. This classification encompasses a broad range of habitat types, including high marsh transition zone, riparian forest, non-tidal brackish marsh, valley freshwater marsh, and other estuarine and palustrine habitats. Extensive freshwater/brackish wetlands occurred in the river valleys of each lagoon upslope of

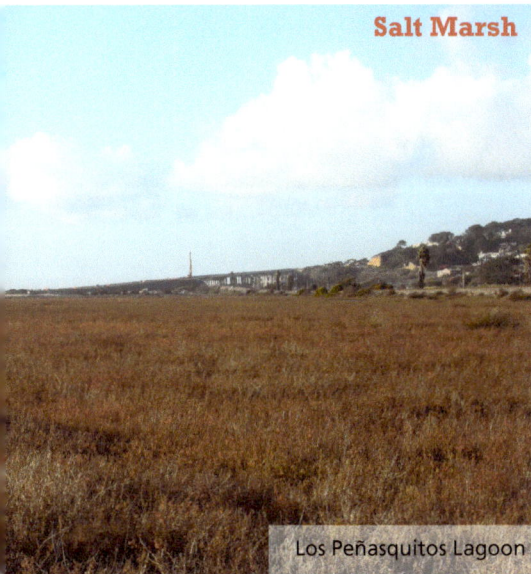
Salt Marsh

Los Peñasquitos Lagoon

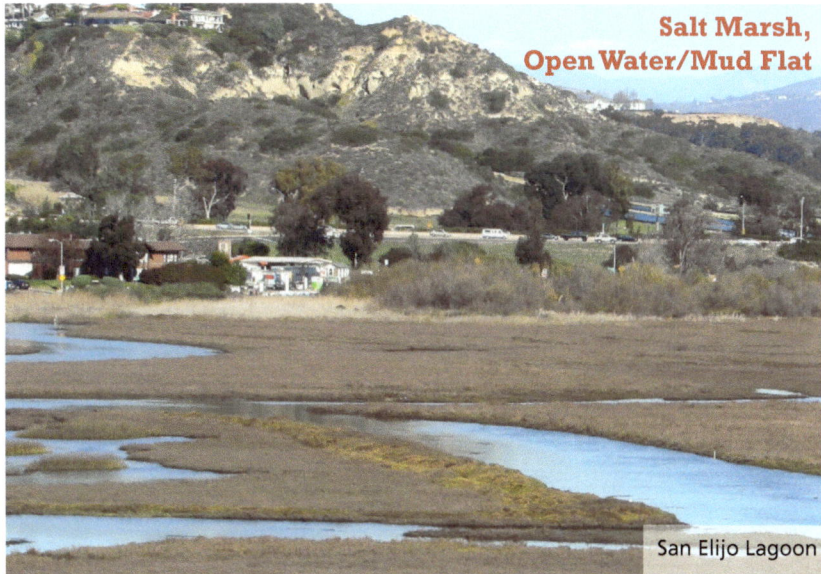
Salt Marsh, Open Water/Mud Flat

San Elijo Lagoon

Open Water and Mud Flat

San Elijo Lagoon

Salt Flat (Seasonally Flooded)

San Elijo Lagoon

Salt Flat (Seasonally Flooded)

Agua Hedionda Lagoon

Freshwater/Brackish Marsh

San Dieguito Lagoon

estuarine habitat types. Common plants within the freshwater/brackish wetlands, especially near the margins of the lagoons, likely included cattails (*Typha* spp.), sedges (e.g., *Cyperus* spp.), rushes (*Juncus* spp.), tules (*Schoenoplectus acutus*), and saltbush (*Atriplex* spp.). Along creeks, common riparian species likely included western sycamore (*Platanus racemosa*), willow (*Salix* spp.), and oaks (*Quercus* spp.). In many cases, the transition between fresh, brackish, and saline habitats would have been gradual rather than abrupt, and would have varied from year to year (Purer 1942).

3. REGIONAL CONTEXT

An understanding of the historical and geophysical context within which the lagoons formed and evolved is an important part of reconstructing the historical characteristics of northern San Diego County's estuaries. Many fundamental aspects of the lagoons, including their size, habitat type distribution, and degree of seasonal and interannual variability, were a direct consequence of their environmental setting and geologic history. Regional climate, tidal dynamics, watershed hydrology, and sediment dynamics all interacted to shape and reshape the systems over time. In addition to these physical processes and controls, human land and water uses have profoundly affected each estuary over the past two centuries. Understanding the timing and nature of these changes is critical to accurately interpreting historical data. This chapter provides a broad overview of the environmental and historical context relevant to understanding the historical ecology of these estuaries.

Environmental Setting

Regional Climate

San Diego County experiences a two-season Mediterranean climate characterized by warm, dry summers and cool, wet winters; over 70% of rainfall occurrs on average between the months of December and March (County of San Diego Department of Public Works 2003). Mean annual precipitation is just under 10 inches along the coast and over 30 inches in the mountains to the east. Annual and decadal wet/dry cycles are driven by large-scale climate phenomena including the El Niño Southern Oscillation (ENSO) and the Pacific Decadal Oscillation (PDO). A recent analysis of southern California streamflow records shows that over the past century, large storm-induced flood flows are much more frequent during ENSO years than non-ENSO years (Andrews et al. 2004). The cycles of droughts and floods documented in 18th and 19th centuries were influenced by these same climatic drivers.

Tidal Dynamics

The region experiences a mixed, semi-diurnal tide regime, meaning that there are usually two high tides (a high and higher high) and two low tides (a low and lower low) per day. The tide gage in La

(top) San Luis Rey Mission, ca. 1895. (photo #81:9922, courtesy of San Diego History Center)

Regional precipitation record, 1774-2012. Values represent annual rainfall for an October through September water year. Precipitation data for 1851-2012 was obtained from meteorological records from San Diego Airport (dark green bars; National Weather Service 2013, Western Regional Climate Center 2013). Precipitation estimates for 1774-1834 were calculated using Rowntree's (1985) rainfall index for southern California, which was constructed from crop harvest records from southern California missions. Precipitation estimates for 1835-1850 were derived from Lynch's (1931) rainfall index for the San Diego area, which for these years is based on historical diary entries describing weather conditions. Rainfall indices were translated into precipitation estimates (light green bars) using a mean annual rainfall value of 9.92 inches for the base period 1851-2005. Lower certainty is ascribed to precipitation estimates derived from the rainfall indices (particularly for the 1835-1850 period) than to the subsequent meteorological records.

1809
Ruinous drought.

—JAMES 1911

1821
A remarkable flood occurred in San Diego during September and October, 1821, causing extensive damage.

—KUHN AND SHEPARD 1984

1869-77
The season of least rain was that of 1876-1877, when 3.75 inches were recorded…The driest period ever known extended from 1869 to 1872, when the rainfall averaged 4.5 inches a season.

—CARPENTER 1913

1883-4
Season of unusual rainfall.

—HALL 1888

1884
The flat at Cordero [Peñasquitos] was entirely submerged at last accounts and the amount of damage to the road there cannot be ascertained. The same may be said of all the country adjacent to those rivers that are crossed by the road.

—SAN DIEGO UNION IN LOS ANGELES HERALD 3/20/1884

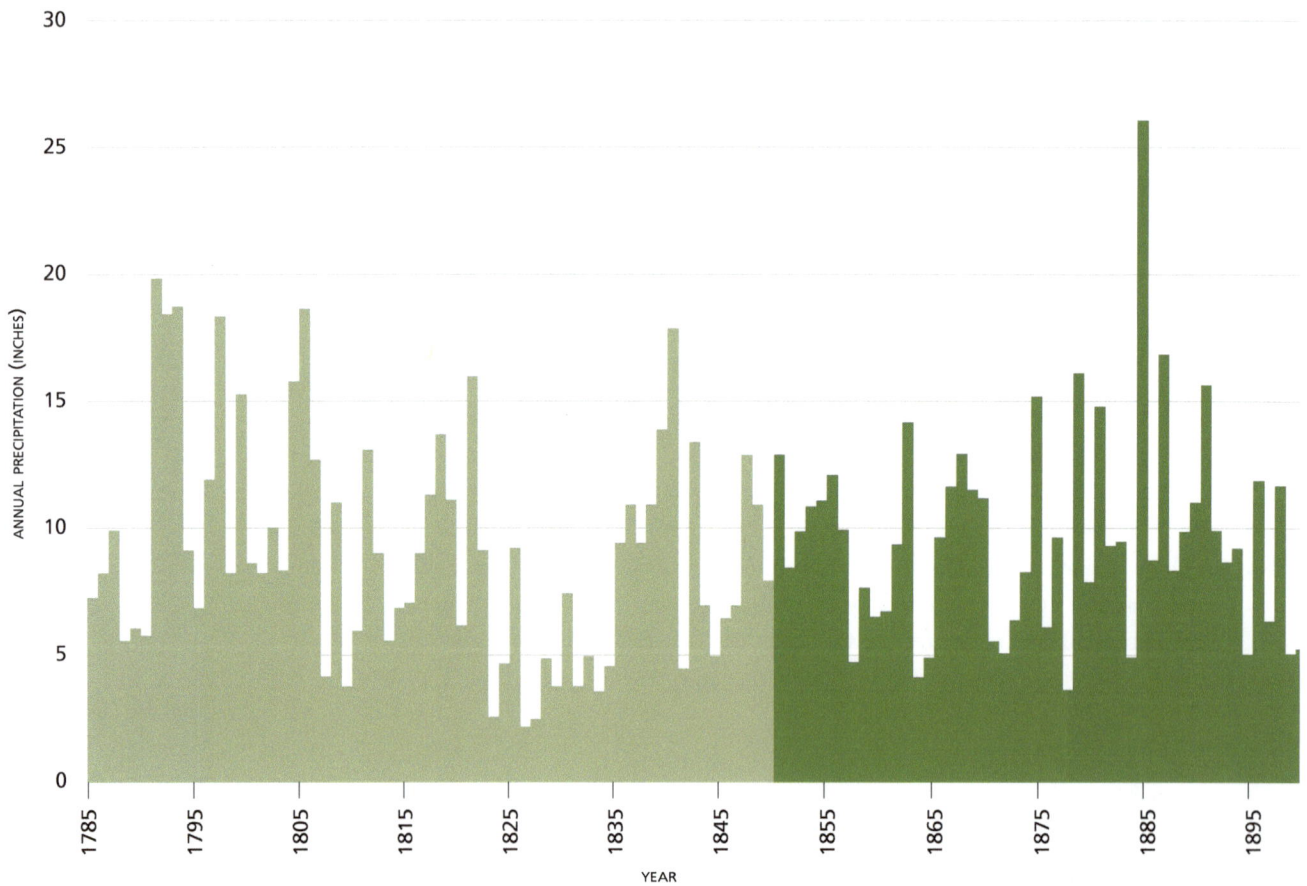

1889

It all broke with the big rains of December 1889. They came and would not stop. All over the county, railroad tracks were washed out, as were roads and bridges. Crops were ruined and livestock drowned. Del Mar was practically washed out to sea. Its dirt streets were rivers of mud, especially the road to the beach....By Christmas the town was isolated. Roads to the north and south were under water in many places, and the railroad had lost miles of track and most of its bridges along the line.

—EWING 1988

1898-1905

Ground water was first largely used in San Diego County in 1898—the first year of serious drought subsequent to the early eighties, when the extensive settlement of the county was begun. This drought continued with varying severity until 1905, rainfall and stream flow being far below normal throughout the whole period.

—ELLIS AND LEE 1919

1905

Rain continues to fall in showers. Last night .40 of an inch was added to the fall, accompanied by high wind. This morning a heavy shower continued for two hours. For the storm the figures are 4.23; for the season, 9.02, an excess of 3.21 over the fifty-year normal. All the big reservoirs in the county are being filled.

—SAN FRANCISCO CALL 2/7 1905

1916

For a week San Diego County has been struggling in the grip of a storm and floods that have extended all over the state and that for severity and in point of damage done exceed anything experienced in years past.

—OCEANSIDE BLADE 1916 IN USACE 1973

Calculated from rainfall index

Obtained from meteorological records

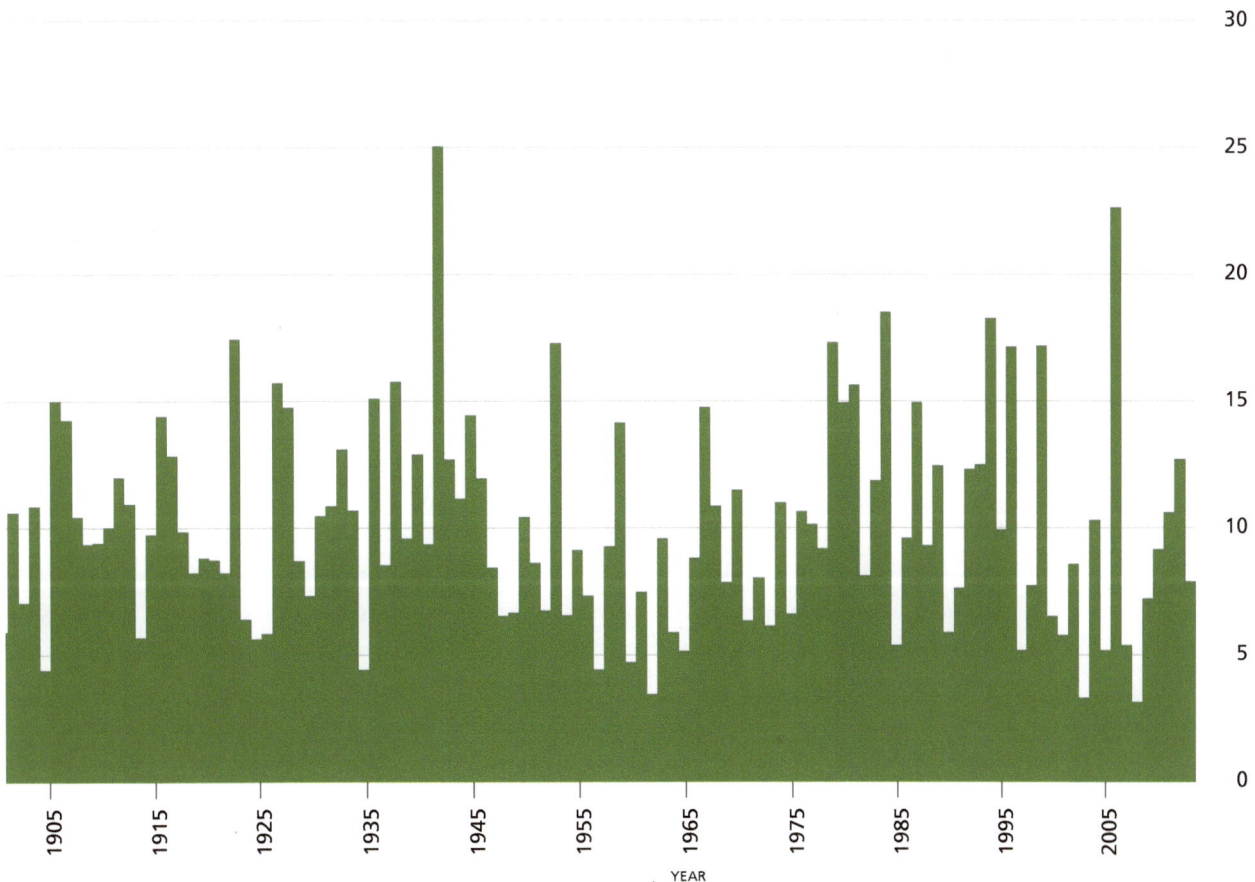

YEAR

Jolla, CA (NOAA station 9410230) shows a long-term mean tidal range (difference between Mean High Water [MHW] and Mean Low Water [MLW]) of 3.69 feet and a mean diurnal tidal range (difference between Mean Higher High Water [MHHW] and Mean Lower Low Water [MLLW] of 5.30 feet. In general, storm-induced increases in tidal elevation are relatively small (typically less than 2 feet above the MHHW elevation) when compared to normal tidal fluctuations.

Lagoon Formation and Evolution

The evolution of the region's coastal lagoons has been largely governed by the dynamic interplay of sedimentation from the watershed and sea level rise. North County lagoons were formed over the course of the current interglacial period as a series of coastal valleys, which flooded and then became filled in with sediment (Masters and Aiello 2007, Jacobs et al. 2010). Rapid sea level rise after the Last Glacial Maximum (about 18,000 years ago) flooded the incised valleys cut when sea levels were low during the late Pleistocene and early Holocene, forming deep, open embayments (Masters and Gallegos 1997, Masters and Aiello 2007). As sea level rise slowed during the middle Holocene (about 5,000-6,000 years ago), wave action along the coast caused cobble spits to form at the mouths of these embayments, and estuaries began to fill in with sediment. By the late Holocene, the embayments had become shallow, intermittently tidal coastal lagoons (Masters and Aiello 2007).

Though North County lagoons have gradually transitioned from deep embayments to shallow lagoons over the past 6,000 years, this progression has not been strictly linear. Data taken from cores indicate alterations between periods of more continuous tidal influence and periods with more intermittently tidal conditions during this time. This has been documented for both Batiquitos Lagoon (e.g., Phillips et al. 1978) and San Elijo Lagoon (Byrd n.d.). These cycles likely reflect extreme climate events such as megadroughts and megafloods (Masters and Aiello 2007); for example, a large flood approximately 1,500 years ago may have created more open conditions (Gallegos 2002).

Watershed Hydrology

Each lagoon receives freshwater input from its watershed through runoff and subsurface flow, as well as through the stream channels that drain into each estuary. The majority of the six lagoons studied have relatively small watersheds, ranging from approximately 22 to 94 square miles (the exception is San Dieguito Lagoon, whose watershed is just under 350 square miles; see graph on facing page).

Historically, the streamflow entering each lagoon was highly seasonally variable, with the bulk of freshwater inputs occurring during the wet season and little surface flow reaching the lagoons during much of the dry season. Most of the region's creeks had extensive intermittent reaches, particularly as they ran through broad alluvial valleys. However, the historical existence of extensive freshwater/brackish wetlands at the upslope margins of each lagoon, formed in high-groundwater areas where creeks spread into wetland complexes above each estuary, reflects the presence of at least some diffuse perennial freshwater inputs to each lagoon in the form of groundwater and surface runoff.

Natural watershed hydrology has been significantly altered by a variety of land and water uses. Early modifications, such as dam construction and surface and groundwater diversions in the 19th and early 20th centuries, likely decreased overall freshwater inputs to the lagoons. In contrast, urban runoff, irrigation, and wastewater discharge associated with urban development in the mid- to late 20th century has tended to increase freshwater inputs. As a result, today many of the lagoons receive surface flow year-round (Welker and Patton 1995, White and Greer 2006; see page 171 for further discussion of historical streamflow patterns and page 38 for further discussion of land and water use history).

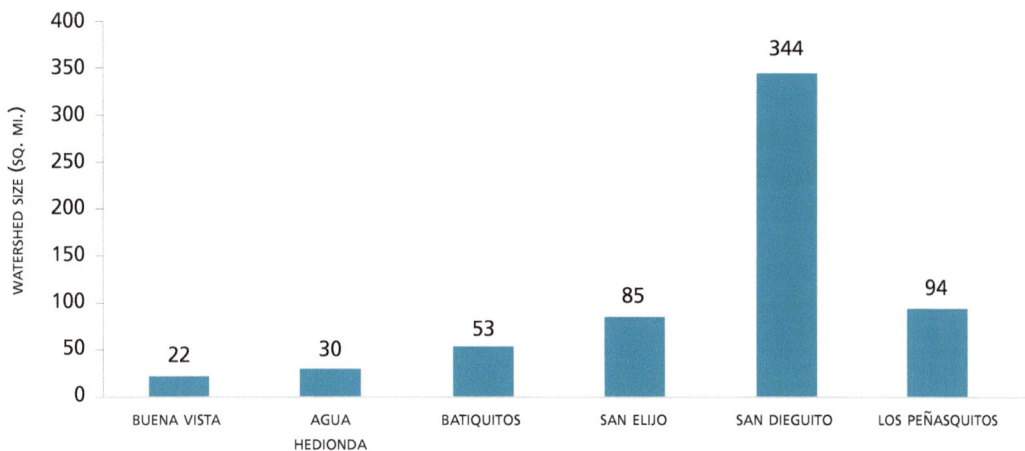

Watershed size for each lagoon. The order of the lagoons from left to right in the chart corresponds to the order of the lagoons north to south along the coast. (data from WyGISC 2008)

Sediment Dynamics

Sediment is supplied to the Oceanside littoral cell from a series of streams stretching from San Juan Creek (Orange County) in the north to Los Peñasquitos Creek to the south, draining a total watershed area of 2,100 square miles (Orme et al. 2011). These streams drain watersheds characterized by competent, plutonic igneous bedrock (e.g., gabbro and tonalite) in high elevation, headwater catchments and more erosive marine sedimentary bedrock (e.g,, fine-grain sandstone and mudstone) near stream mouths and along eroding coastal bluffs (Jennings et al 1977). Regional tectonic activity has resulted in localized sheared and weakened igneous and sedimentary bedrock units that contribute both coarse and fine sediment to the channel network. Dominant sediment sources include landslides and hillslope mass wasting features triggered by large storm events and earthquakes and often exacerbated by wildfire.

Sediment transport through the channel network and out to the Oceanside littoral cell is very episodic, with wet period sediment flux about 15 times greater than in the dry period (Young and Ashford 2006). Average annual sediment delivery to the cell from streams has been estimated to be approximately 590,000 cubic yards under natural conditions (Nordstrom and Inman 1973), or approximately 250,000 to 350,000 cubic yards when accounting for in-channel sediment storage (Nordstrom and Inman 1973, Patsch and Griggs 2006). Over the past century, watershed modifications such as large water-supply dams and other sediment impoundments have resulted in a relative decrease in fluvial sediment supply to the littoral cell and a relative increase in the contribution from eroding coastal bluffs (Inman and Masters 1991). Within the San Dieguito River watershed, for example, studies assessing average annual coarse sediment (gravel and sand) yield from the watershed show an order of magnitude yield decrease from pre- to post-dam conditions ($\sim$100 tkm^{2}yr^{-1} to $\sim$10 tkm^{2}yr^{-1}; Brownlie and Taylor 1981). These changes were likely also accompanied by shifts in dominant sediment size delivered to the marsh.

Over the past several decades, annual sediment transport patterns within the Oceanside littoral cell have varied considerably. Beginning in the late 1970s, the average annual longshore transport direction changed from southerly to a more even mix of northerly and southerly transport following a shift in ocean currents. This change in the direction of longshore sediment transport has apparently resulted in a more uneven distribution of the sand supplied to coastal beaches, which may also have implications for lagoons inlet closure dynamics (Flick 2005). Littoral sediment transport dynamics also vary on a seasonal and interannual basis: during ENSO years, for example, winter westerly and southwesterly waves mute the southern transport, while summer southern swells may emphasize more northerly transport (Patsch and Griggs 2006).

Land and Water Use History

Humans have altered the northern San Diego County landscape in a variety of ways, resulting in a wide range of impacts to the coastal lagoons and their watersheds. Some activities, such as the filling, excavation, and development of wetland areas, have involved direct modifications to the lagoons. Others, such as water diversion for irrigation or livestock grazing within the lagoon watersheds, have impacted the lagoons indirectly (e.g., by changing patterns of freshwater input or sediment delivery to the lagoons). Because humans have been living along the northern San Diego County coast for thousands of years, none of the historical sources consulted in this study depict a landscape entirely free from human influence, though the signature of human activity is much more prominent in later sources. Accurate interpretation of lagoon characteristics from the historical record thus requires an understanding of the prevailing land uses affecting the lagoons at the different points in time.

In addition to providing critical context for the interpretation of historical ecological conditions, information about past land uses can also provide clues to physical characteristics of the historical landscape. For example, a historical map that shows a road crossing a salt flat is an indication that the lagoon was passable to vehicles for at least some portions of the year, and was therefore seasonally dry.

This section describes some of the most significant land use changes that have occurred within the study area over the past several centuries, with a focus on impacts since the arrival of Euro-American settlers. Discussion of specific land use changes pertaining to individual lagoons is included in chapters four through nine.

Prehistoric Cultural Context

Humans have lived along the San Diego County coast since at least 9,000 years ago (Gallegos 1992, Gallegos 2002). At the time of European contact in 1769, the San Diego County coastline was occupied by two tribes, the Kumeyaay (also referred to as Diegueño) and the Luiseño.

The Kumeyaay occupied a large area extending from Ensenada, Mexico north beyond Batiquitos Lagoon. To the north of the Kumeyaay, Luiseño territory extended from Agua Hedionda Lagoon north to Riverside County (Iversen et al. 2009). Estimates of Kumeyaay population size range from 10,000 to 20,000 (Gallegos 2002). Though the Luiseño population size was only an estimated 5,000 to 10,000, they occupied a much smaller area than the Kumeyaay, and thus population densities were significantly higher (Byrd and Reddy 2002). Traveling through northern San Diego County in July 1769, explorer Juan Crespí recorded encountering many Kumeyaay and Luiseño settlements, including numerous villages along the coast in valleys near the lagoons (Crespí and Brown 2001).

Land Use Timeline: Northern San Diego County

1769	The Portolá expedition arrives in San Diego
1769	Mission San Diego de Alcalá established
1798	Mission San Luis Rey established
1820s-30s	Peak of mission-era grazing
1834	Secularization of the missions
1881-82	California Southern Railroad constructed
Late 1800s/early 1900s	Rapid expansion of agriculture
1910s	Population of San Diego County reaches 100,000
1910s	Beginning of extensive groundwater pumping and irrigation
1912-15	Pacific Coast Highway constructed
1920-70s	Treated wastewater discharged into various lagoons
1965	Interstate 5 Highway constructed
Late 20th century	Rapid urbanization; implementation of conservation and restoration measures

For thousands of years, early inhabitants used the deep, open embayments formerly found where the estuaries are today as sources of fish, shellfish, and other dietary staples (Masters and Gallegos 1997; see page 36 for more information on the transformation of these coastal embayments to shallow lagoons). Early archaeological research hypothesized that these changing environmental conditions heavily influenced patterns of human occupation on the southern California coast, triggering a major migration away from the coast towards more inland areas (Warren and Pavesic 1963). However, more recent findings show that coastal settlements actually persisted throughout this period (Gallegos 2002, Byrd 2004).

Livestock Grazing

The arrival of Spanish missionaries in the late 18th century displaced native communities and introduced a suite of new land uses. With the establishment of Mission San Diego de Alcalá (founded 1769) and Mission San Luis Rey (founded 1798), livestock grazing became the dominant land use in San Diego County. Cattle and sheep were most abundant, though livestock holdings also included lesser numbers of goats, pigs, horses, and mules (Engelhardt 1920, Engelhardt 1921). Grazing areas for the missions encompassed large tracts of land in the western portion of present-day San Diego County, extending into the watersheds of all six lagoons in the study area (Engelhardt 1920, Bowman 1947).

Livestock at Mission San Diego numbered just 245 in 1773, increasing to about 13,000 by 1800 and peaking at about 30,200 in 1822 (Engelhardt 1920, Bowman 1947; see graph on following page). Grazing lands for Mission San Diego encompassed an estimated 155,000 acres, resulting in an estimated peak stocking rate of approximately 16-17 acres/head for cattle (Bowman 1947). Herds at Mission San Luis Rey were considerably larger than at Mission San Diego, increasing from 800 in 1798 to about 16,800 in 1810 and peaking at

At almost every ranch there would be hanging quarters of fresh beef. When we asked to buy they would hand us a knife and tell us to cut all we wanted. They were glad to have us take meat in that way, rather than we should shoot down a steer, as travelers generally did, because then they lost the hide which was the main value of the animal. We passed places where apparently thousands of cattle had been killed, and were told that all but the hides and horns were thrown away.

—SMITH 1849, TRAVELING FROM SAN DIEGO TO SAN LUIS REY EN ROUTE TO SAN FRANCISCO

about 58,800 (comprised of roughly 50% cattle and 50% sheep) in 1828 (Engelhardt 1921, Bowman 1947). Grazing lands for the mission encompassed an estimated 110,700 acres, resulting in a peak cattle stocking rate of approximately 4 acres/head (Bowman 1947).

Both San Diego and San Luis Rey's grazing lands extended well beyond these six watersheds: roughly half of Mission San Diego's grazing lands were within the San Dieguito and Los Peñasquitos watersheds, while only about one-fifth of Mission San Luis Rey's grazing lands were found within the watersheds of Buena Vista, Agua Hedionda, Batiquitos, and San Elijo lagoons (Bowman 1947). In the case of Mission San Luis Rey, Father Antonio Peyri cited "the lack of water and pastures" as "the reason for having [livestock] so scattered" (Peyri 1827 in Engelhardt 1921). Because the mission grazing lands were so spread out, only a portion of each mission's livestock were grazed within the lagoons' watersheds. Furthermore, only a portion of each watershed was grazed, ranging from the majority of the watershed area (Agua Hedionda Lagoon, ~65% of area grazed) to almost none of the area (San Elijo Lagoon, <1% of area grazed; Bowman 1947). Taking both of these factors into account, average peak mission cattle stocking rates across the study area are estimated to have been 6-22 acres/head in the Buena Vista, Agua Hedionda, and Batiquitos lagoon watersheds, about 60 acres/head in the San Dieguito Lagoon watershed, and about 270 acres/head in the Los Peñasquitos Lagoon watershed; cattle stocking rates in the San Elijo Lagoon watershed were negligible. As a relevant side note, the Mexican cattle ubiquitously raised by the Mission and early ranchos were smaller and required less forage than American cattle, which were only introduced to southern California by the 1860s (Adams 1946, Burcham 1956). Sheep required only about one-fifth the amount of forage as Mexican cattle (Lightner 2013).

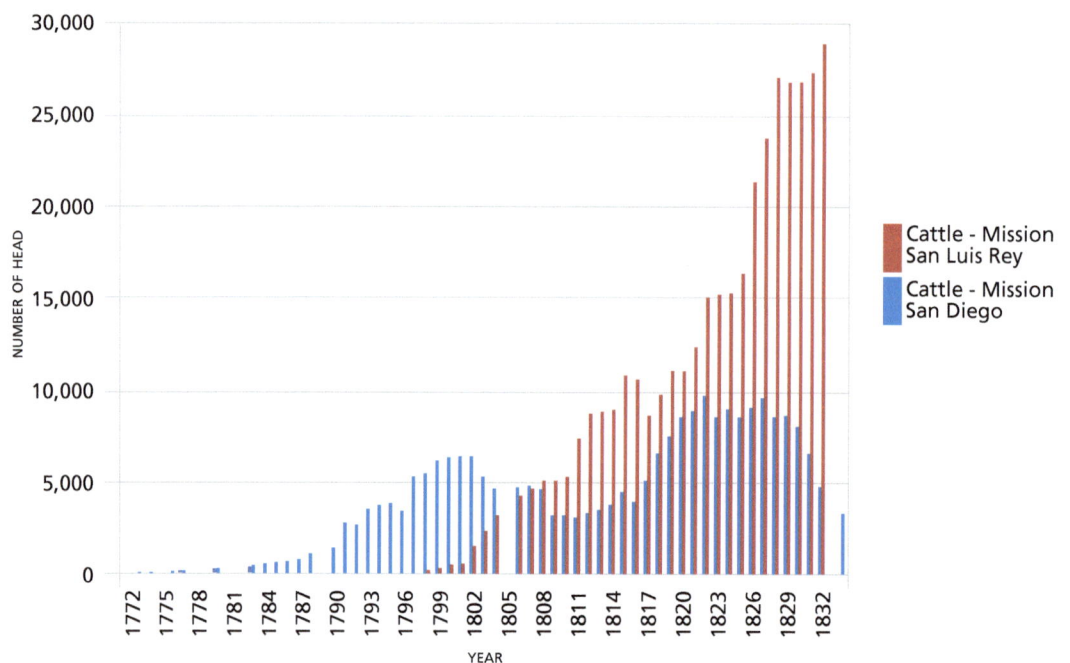

Mission-era cattle stocking levels at Mission San Diego and Mission San Luis Rey. After several decades of low stocking levels in the late 18th and early 19th centuries, livestock numbers peaked in the 1820s and 30s, but subsequently crashed following the collapse of the mission system in 1834. (data from Engelhardt 1920, 1921)

After the collapse of the mission system in 1834, ranching continued on private land grants through-out the county, though the herds were far smaller than those at the peak of the mission era (Hughes 1975). Few data are available on stocking rates for individual ranchos in northern San Diego County during this time, but it appears that holdings were not particularly extensive. The first land grant in the county, Rancho Los Peñasquitos (which was granted in 1823 and and occupied 14% of Los Peñasquitos Lagoon watershed), reportedly had just "50 cattle, 20 horses, and 8 mules" in 1828 (Smythe 1908). An 1850 tax assessment of Rancho Agua Hedionda (which was granted in 1842 and occupied approximately 60% of Agua Hedionda Lagoon's watershed) lists 1,000 untamed cattle along with smaller numbers of other animals (Christenson and Sweet 2008).

Cattle ranching throughout much of southern California experienced an unprecedented boom in the early 1850s. Prior to 1848, tallow (rendered fat) and hides were the main commodities derived from cattle ranching, but beef production rapidly overtook these industries in importance as the Gold Rush created a vast new market for meat (Cleland [1941]1990). Northern San Diego County ranchers bene-fited from the new beef market to some extent, but the region's arid rangeland and distance from major trading ports limited ranchers' ability to capitalize on the Gold Rush cattle boom (Hughes 1975).

Census data for San Diego County show just 5,164 head of cattle in 1852 and 15,452 head in 1860, compared with 71,078 head of cattle in Los Angeles County in 1860 (U.S. Census 1853, 1864; see graph below). Sheep were also relatively scarce: San Diego County had just 13,768 head in 1860, compared with 94,639 head in Los Angeles County. These stock were also distributed over a large area, since San Diego County was more than three times larger in the mid-19th century than it is today.

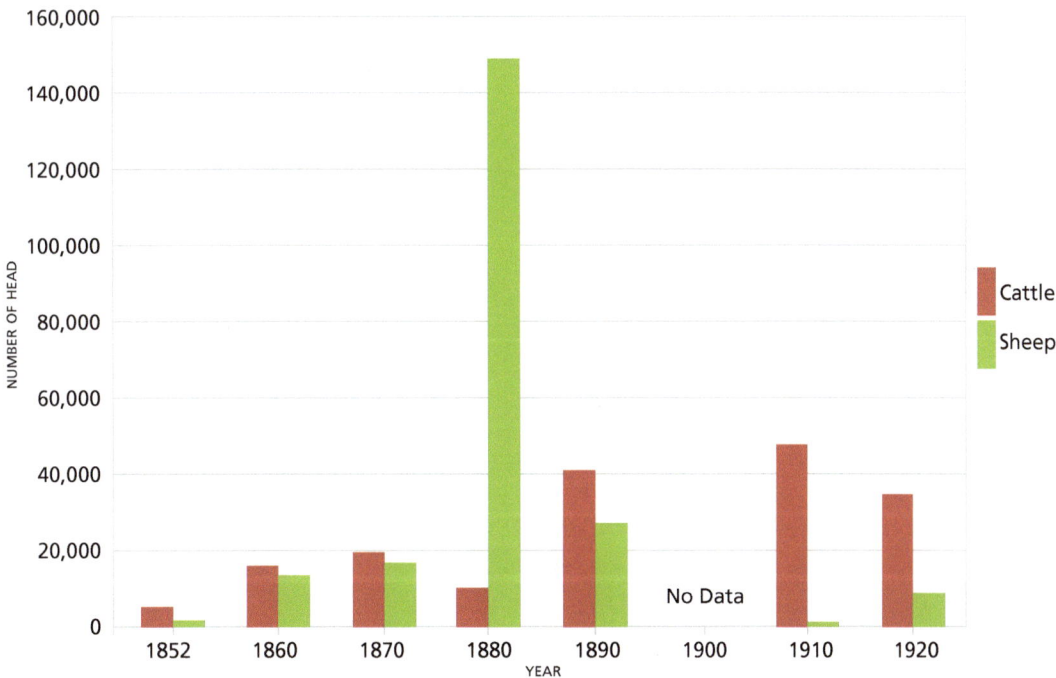

Cattle and sheep in San Diego County, 1852-1920. Livestock numbers remained relatively low throughout much of the 19th century. Sheep ranching declined rapidly following a brief peak around 1880, while cattle ranching increased in the late 19th and early 20th centuries. (data from U.S. Census 1853, 1864, 1872, 1882, 1895; Wade et al. 2009)

Additional hardships continued to plague cattle ranchers during the 1860s and 70s. Droughts in the late 1850s and 1863-64 caused thousands of cattle throughout southern California to die of starvation or be slaughtered in anticipation of the lack of food and water, while flooding in 1861-62 drowned many more (Cleland [1941]1990, Wade et al. 2009). Disease, competition from Midwestern cattle ranchers, and the rise of the sheep industry also took a heavy toll on southern California's cattle industry (Cleland [1941]1990, Wade et al. 2009). By the late 19th and early 20th centuries, however, San Diego County's cattle industry had recovered, while sheep ranching had largely succumbed to drought and disease (Cleland [1941]1990).

Grazing has the potential to result in a variety of ecological impacts, including soil erosion, soil compaction, reduced rainfall infiltration, increased runoff, reduced water quality, decreased plant biomass, and changes in plant species composition (Burcham 1961, Trimble 1995, Bilotta 2007). Given the lack of precise data on historical grazing practices, post-mission era stocking rates, and the distribution of livestock throughout northern San Diego County's rangelands, it is only possible to speculate about the impacts of 19th century grazing on the county's coastal lagoons. Though the possibility of significant grazing impacts during the mission and rancho eras cannot be discounted, the data discussed here indicate that impacts might not have been as pervasive or prolonged as in more heavily grazed regions of coastal California. Grazing densities were relatively low for much of the 18th and 19th centuries, with the highest stocking rates occurring in the 1820s-30s and 1880s-90s. Grazing pressure also varied spatially: overall livestock density appears to have been substantially higher within the rangelands of Mission San Luis Rey than within those of Mission San Diego, and certain watersheds experienced more extensive grazing than others. In the Agua Hedionda Lagoon watershed, for example, relatively large areas were used for grazing during both the mission and rancho eras, while in the San Elijo Lagoon watershed there appears to have been very little grazing as a result of its relatively remote location far from both missions. This hypothesis is further supported by the earliest observations of the lagoons, which predate the establishment of the missions and the onset of any stock grazing and describe conditions broadly consistent (including the presence of salt flats and freshwater/brackish wetlands) with those described by later sources (Crespí and Brown 2001).

Agriculture

In addition to livestock grazing, agriculture was the other noteworthy early land use. The aridity and isolation of northern San Diego County, as well as larger-scale economic conditions, prevented agriculture from flourishing in the region until relatively late in comparison to neighboring areas (W.W. Elliott & Co. 1883, Rodgers 1889, Heilbron 1936, Roth and Associates 1990, Flannigan et al. 1993). For example, though the climate was well-suited to growing fruit, the challenges of preserving the harvest long enough to transport it to the larger markets to the north at first discouraged the establishment of this industry. (One account described a "fear of raising fruit that could not be given away"; W.W. Elliott & Co. 1883). The T-sheets provide a glimpse of the relatively modest extent of farmland in the immediate vicinity of the lagoons as late as the 1880s (at left). Small patches of row crops and orchards, up to several dozen acres in size, are shown scattered around most of the lagoons in valleys and on the coastal bluffs.

By around the turn of the century, however, agriculture had assumed a dominant role in San Diego County's economy (see graph on facing page). Wheat, wool, and honey were among the earliest economically important crops grown in San Diego County, and by 1886 they

Patches of farmland in the San Dieguito Valley, shown as dashed black lines on the T-sheet from 1889. (Rodgers and Nelson 1889; courtesy of NOAA)

My apology for not incorporating with this report the various statistical data relative to the number and quantity of cattle, swine, beans, cabbage, goats, buckwheat, hay, hemp, barley, oats, onions, cheese, turnips, eggs, butter and beeswax, as contemplated by the law, as given in your circular, must be my very limited interest in matters so entirely unconnected with my profession and confessed ignorance and want of information upon the subjects indicated, not to mention the great and unremunerated expense inseparable upon the performance of the duty. My general impression of the live stock department is, that several thousand cattle, of a fierce and savage breed, infest the valleys of this whole county, making the Surveyor's duty of running lines through their range, a matter of some personal risk and uncomfortable foreboding (I had an unsuspecting flagman prostrated once by a charge in the rear from an infuriated bull) that swine are not numerous…a similar remark being applicable to sheep and goats. …The number of rivers in this county, according to the map, is not far from correct; but their locations and courses do violence to all notions of topographical propriety.

—SAN DIEGO COUNTY SURVEYOR CHARLES H. POOLE, 1855

At present the most serious trouble in San Diego County seems to be the want of a home market at which grapes, apricots and other fruits can be immediately and profitably disposed of without drying or any other preserving labor on the part of the producer; in short, canneries, wineries, distilleries, etc. San Diego has been so isolated that nothing could be profitably shipped to the canneries of Los Angeles or San Bernardino Counties.

—W.W. ELLIOTT & CO. 1883, SAN BERNARDINO COUNTY

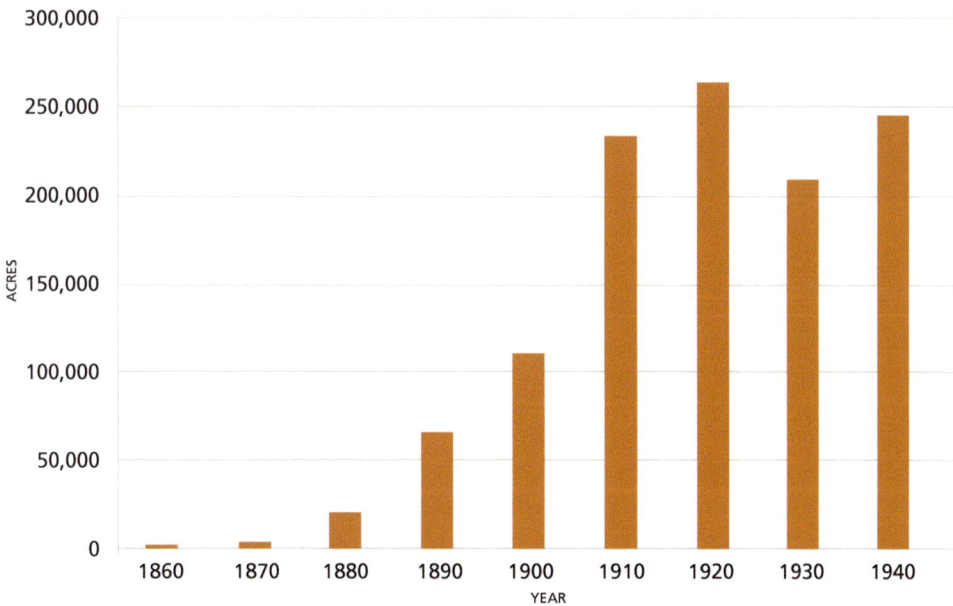

U.S. Census data for cropland acreage in San Diego County from 1860-1940, showing an expansion of cropland in the late 19th and early 20th centuries. Our analysis is based on the USGS Cropland by County dataset, which processed the census data to account for changes in census terminology and county boundaries over time (Bliss 2002; see Waisanen and Bliss 2002 for processing methodology).

"Shipping sugar beets near Oceanside," ca. 1913. (Unknown ca. 1913; courtesy of Mandeville Special Collections, UC San Diego)

represented the "leading products of the County in point of value" (Gunn 1886). Other early agricultural products included barley, corn, alfalfa, dairy products, fruit, lima beans, sugar beets, and oats (U.S. Census 1864, 1872, 1882; Gunn 1886; California Development Board et al. 1923). Avocados and flowers became important crops in the northern portion of the study area in the 1920s (California Development Board et al. 1923, Storie and Carpenter 1929b).

Railroads and Highways

Prior to the 1880s, the absence of transportation infrastructure hindered travel through northern San Diego County and kept the area relatively isolated from population centers in San Diego and Los Angeles (W.W. Elliott & Co. 1883). The construction of the California Southern Railroad between 1881 and 1885 made the region accessible to larger numbers of people for the first time, eventually precipitating substantial changes in land use. The railroad was chartered in 1880 with the intention of providing a rail connection between San Diego and the national line to the north (Lowell 1985). Construction began in late 1880 in National City (south of San Diego) and progressed rapidly through 1881 (Hoyt 1954). By January 1882, the line between National City and Oceanside was operational, but it was not until 1885 that a connection was established between the California Southern and the Santa Fe Railroad's transcontinental line in Barstow (Hoyt 1954, Dodge 1958).

The railroad line ran along the coast in the northern part of the county, crossing all six lagoons. With the exception of Los Peñasquitos, the line crossed the lagoons quite near (within ~500 to 1,500 feet) the coast; at Los Peñasquitos the line ran approximately one mile inland. The crossings consisted of elevated berms interrupted by short (~60-520 foot) bridges.

Floods periodically damaged large sections of the track (Lowell 1985, Flannigan et al. 1993, Hawthorne 2003). In December 1889, for instance, just seven years after track had been laid between National City and Oceanside, the *Los Angeles Herald* reported:

Made an excursion over the road to-day to the end of the track at San Dieguito, 20 miles from town. ...Track laying is now going forward at the rate of about a mile a day, and it is expected that the track will reach San Luis Rey in ten days or a fortnight.

—LOS ANGELES HERALD
11/13/1881

A large force of men are at work night and day repairing the washouts on the California Southern road. At the mouth of the San Dieguito valley, in the lowland, there are five bridges washed out of position, and a full half mile of track has been twisted out of place. (*Los Angeles Herald* 1889)

The railroad line through north San Diego County is still in use today, though portions of it have been re-aligned.

An informal route for travelers on foot or horseback existed along the northern San Diego County coast at least as early as the 1840s. By the mid-1880s (as seen on the T-sheets), a crude road extended along the coast. Without bridges spanning the coastal lagoons and rivers, however, the road was often impassable in wet weather or at high tide. In 1906, for instance, vehicles had to "go body deep in mud for a long distance" at Los Peñasquitos Lagoon (Oceanside Blade 3/24/1906 in Hawthorne 2003). Construction of the Coast Highway between Oceanside and San Diego proceeded over a period of several years, beginning in 1912, and the entire route was completed by June 1915 (Hawthorne 2003). The highway was widened in the 1920s, and various portions have subsequently been re-aligned or rebuilt. In 1925 the road was designated a part of Highway 101. Interstate 5, constructed between 1965 and 1967, has a much wider footprint than the Coast Highway. The highway crosses each lagoon (with the exception of Los Peñasquitos) to the east of the Coast Highway and the railroad. Plans to widen I-5 are currently in development (Caltrans and SANDAG 2013).

The railroad berms of the 1880s and subsequent road infrastructure would have impacted the hydrogeomorphic functioning of the lagoons by laterally confining lagoon inlets (and thus restricting inlet migration and altering inlet closure dynamics), changing scour patterns, altering sediment transport and deposition, and redirecting flow (see Stanbro 1971, Meyer 1980, Nordby and Zedler 1991, Goodwin et al. 1992; see also Chapter 11).

Dredging, Filling, and Inlet Modifications

In addition to the construction of the California Southern Railroad, Coast Highway, and Interstate 5 corridors, a variety of other modifications have directly shaped the lagoons as we know them today. The most significant of these activities include dredging, filling, and inlet modifications.

Efforts have been made to manipulate the mouths of all six lagoons at various times in the past, in particular over the last century. Most of these efforts have involved breaching the inlets to establish or maintain tidal connections between the lagoons and the ocean. (The primary exception is at Buena Vista Lagoon, where a weir was installed in the 1940s to exclude tidal influence.) In some cases, permanent infrastructure was installed: at Agua Hedionda, for instance, jetties were constructed in 1954 in order to keep the mouth permanently open to the ocean. Active manipulation of lagoon inlets continues today as a component of contemporary management activities.

Dredging activities have extended well beyond the inlets to include larger areas of some lagoons. Much of Agua Hedionda Lagoon was dredged in the 1950s, for instance. The 1996 Batiquitos Lagoon Enhancement Project likewise involved dredging much of the inland area of the lagoon.

The city fathers of San Diego recently looked over the Sorrento slough [Los Peñasquitos] in company with Mr. Grimm and have promised to move a portion of the road to higher ground by the Torrey pines and to build a bridge so that the Sorrento bad spot will be put in excellent shape for autos or farm wagons or anything else on wheels.
—OCEANSIDE BLADE 11/24/1906 IN HAWTHORNE 2003

J.C. Hayes asked…by what authority the city recently opened the mouth of the lagoon and released the water into the ocean.
—OCEANSIDE BLADE 4/27/1912, DESCRIBING BUENA VISTA LAGOON

Today, instead of a wide expanse of water, Buena Vista lake is a mud flat. Yesterday…a small trench was cut to allow some of the water to escape, but before anything could be done about it the water got away from the workman and was rushing madly into the sea, carrying everything before it.
—OCEANSIDE BLADE-TRIBUNE 1931

"Local citizens opening Sorrento Lagoon." Active manipulation of the lagoon inlets has been a regular occurrence over the past century, as seen in this series of photos taken at Los Peñasquitos Lagoon in November 1969. (USA-C1.9 photos #1284-1, 1284-2, 1284-5, courtesy of Scripps Institution of Oceanography Archives, UC San Diego)

Portions of lagoons have also been filled for infrastructure or development projects. For instance, the northern part San Dieguito Lagoon was filled in the 1930s to construct the Del Mar Fairgrounds. In the 1960s-70s, approximately 100 acres of marsh on the eastern end of Buena Vista Lagoon were filled to build a shopping center (Marcus 1989). Dikes and levees were also constructed in many of the lagoons for various purposes, including the creation of ponds for sewage treatment, duck hunting, or salt evaporation (County of San Diego 1996).

Hydromodification

DAMS, DIVERSIONS, AND IRRIGATION Throughout the study area, surface water diversions and groundwater pumping for irrigation, drinking water, or other uses remained fairly minimal until the early 1900s. The lack of a reliable water supply for irrigation limited early attempts at farming in the region, and those crops that were grown were often dry-farmed (Peyri 1822 in Engelhardt 1921, Hall 1888, Holmes and Pendleton 1918). Though there were limited instances of early groundwater withdrawal for agricultural or commercial use, such as the mineral well developed by John Frazier in the 1880s in present-day Carlsbad (Howard-Jones 1984), significant groundwater pumping did not occur until decades later. A 1919 report, for instance, stated that "extensive utilization of the ground waters was begun only a few years ago" in San Diego County (Ellis and Lee 1919). Similarly, though projects such as the East and West San Pasqual Ditches in the San Dieguito watershed began to divert surface water in the late 19th century, significant diversion and use of surface water in the study area did not occur until the construction of dams, particularly Lake Hodges in 1918 (California Department of Public Works 1949). As a result, as late as 1918 extensive irrigation works were not prevalent in the San Diego County region (Holmes and Pendleton 1918).

Early efforts to store and divert surface water in the region focused on the comparatively large San Dieguito and San Luis Rey rivers (Adams et al. 1912, Holmes and Pendleton 1918, California Department of Public Works 1949). The first dam constructed within the drainage area of the lagoons was Lake Wohlford (initially called Bear Valley Dam), built in 1895 on Escondido Creek within the San Elijo Lagoon watershed. The largest and most significant dam within the lagoons' drainage area is Lake Hodges, built in 1918 on the San Dieguito River. Lake Hodges captures runoff from over 300 square miles, representing nearly 90% of San Dieguito Lagoon's watershed. Other notable dams include Lake Sutherland (1954), also within the San Dieguito Lagoon watershed, Lake Dixon (1971) in the San Elijo Lagoon watershed, and Lake San Marcos (1952) in the Batiquitos Lagoon watershed. Reservoirs enabled the withdrawal of much greater quantities of freshwater from the lagoon watersheds, resulting in a reduction in groundwater recharge and likely altering salinity gradients within the estuaries (Bronson 1968, State Coastal Conservancy and City of Del Mar 1979). Reductions in freshwater inputs and peak flows may have also reduced the frequency of inlet opening (Zedler 2001). Sediment retention behind dams likely decreased fluvial sediment inputs to the lagoons as well, though some of this decrease may have been offset by increased rates of erosion caused by other land uses (Inman and Masters 1991).

WASTEWATER DISCHARGE AND STORMWATER INPUTS During the mid-20th century, wastewater from sewage treatment plants was discharged into most of the lagoons or

The existing irrigation ditches in the county are few in number, and irrigate but small areas.

—HALL 1888

The problem of a supply of water for drinking and purposes of irrigation is the controlling one, and the necessity of immediate expenditure to obtain it deters rapid settlement.

—MENDENHALL 1891

into creeks upstream of the lagoons. Discharges significantly augmented freshwater inputs to the lagoons, with discharge rates in some cases reaching two to three million gallons per day (County of San Diego 1974, Goodwin et al. 1992). Wastewater continued to be discharged into some lagoons until the late 1970s (West 2001). The additional freshwater input altered salinity gradients within the lagoons, contributing to an expansion of freshwater/brackish vegetation in areas historically dominated by more salt-tolerant plants (Dailey et al. 1974, Welker and Patton 1995, Greer and Stow 2003, White and Greer 2006). The wastewater also contributed to elevated nutrient concentrations in the lagoons, leading to eutrophication and water quality problems (Bradshaw and Mudie 1972, Dailey et al. 1974, County of San Diego 1979). Wastewater discharge into the lagoons ceased by the late 1970s, but urbanization in the late 20th century resulted in increased stormwater runoff into the lagoons (Welker and Patton 1995, Greer and Stow 2003, San Elijo Lagoon Conservancy 2005, White and Greer 2006).

Urban Development

The recorded population of San Diego County was less than 10,000 until the 1880s (U.S. Census 1882,1895). Euro-American settlers began to move to the northern part of the county in the 1860s and 1870s, establishing homesteads at the present-day locations of Oceanside, Encinitas, and Cardiff (Heilbron 1936, Hawthorne 2003). The completion of the California Southern Railroad in 1882 opened up the county to larger numbers of settlers (Heilbron 1936, Flannigan et al. 1993). Several coastal towns were established during this period, including Oceanside, Carlsbad, Leucadia, Encinitas, and Del Mar, though most were not incorporated until the mid-20th century (Heilbron 1936, Fetzer 2005).

The population of the county jumped to 35,000 by the 1890s, with approximately half of the residents located in the city of San Diego. This population boom was short-lived,

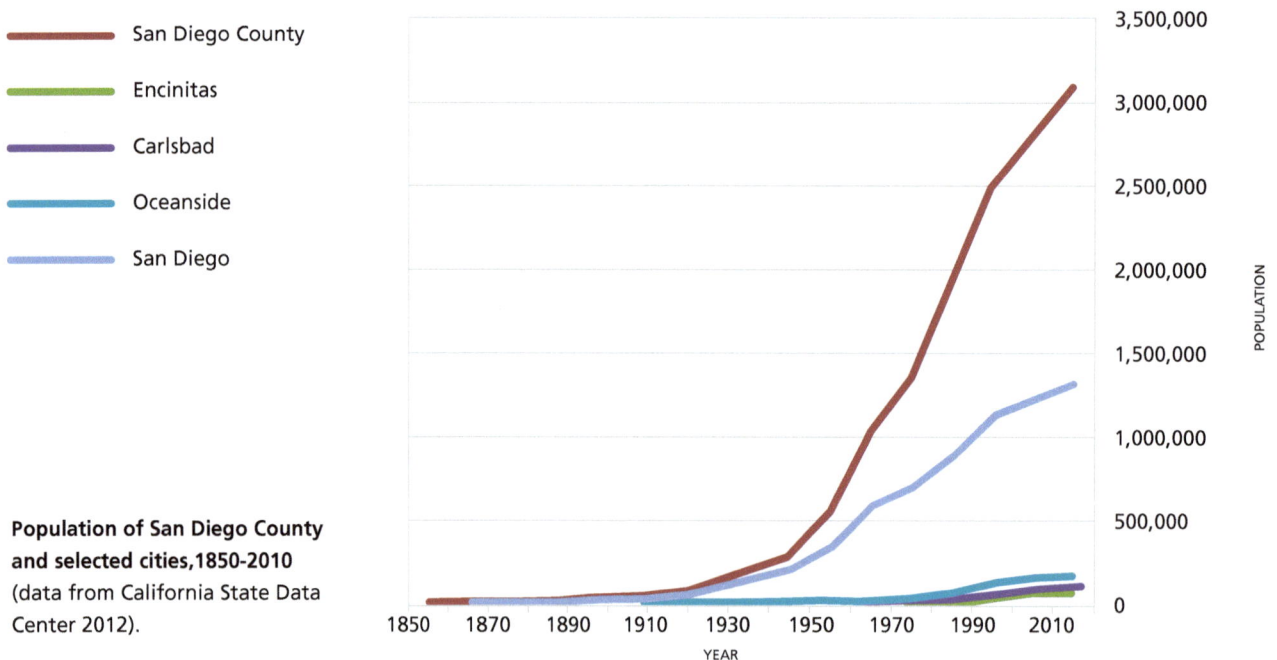

Legend:
- San Diego County
- Encinitas
- Carlsbad
- Oceanside
- San Diego

Population of San Diego County and selected cities,1850-2010 (data from California State Data Center 2012).

however, and by the mid-1890s Carlsbad and other coastal towns were largely abandoned (Helibron 1936). An economic downturn and lack of water for irrigation limited population growth for the next two decades (Roth and Associates 1990).

The 20th century brought public works projects such as the construction of a state highway in 1912-15 and Lake Hodges reservoir in 1918, prompting increased population growth (O'Connell 1987, Hawthorne 2003). New development interests such as the South Coast Land Company, formed in 1906, and the San Dieguito Irrigation District, formed in 1922, brought in many more settlers with the promise of a reliable water supply for agriculture (Heilbron 1936, California Department of Public Works 1949). As a result, the population of San Diego County tripled between 1900 and 1920, and North County cities such as Oceanside also grew rapidly.

Population growth and urban development expanded rapidly in the second half of the 20th century, both in the city of San Diego as well as the northern part of the county. Between 1950 and 2010, for instance, the population of Carlsbad increased from about 4,300 to over 105,300, while the population of Oceanside increased from just under 13,000 to over 167,000. Impacts on the lagoons resulting from this rapid development may include increased dry-season freshwater input, sedimentation, eutrophication, water quality degradation, change in vegetation composition, and loss of wetland habitats (County of San Diego 1970, State Coastal Conservancy and City of Del Mar 1979, Applegate 1985, County of San Diego 1996, White and Greer 2006, Elwany 2011).

Conservation and Restoration Efforts

In recent decades, recognition of the lagoons as important natural resources has given rise to a number of initiatives aimed at protecting and restoring the estuaries. The lagoons have been given a variety of protected area designations, including Ecological Reserve (Buena Vista and San Elijo lagoons), Marine Protected Area (Batiquitos Lagoon), and State Natural Reserve (Los Peñasquitos). Local conservancies and foundations have been established to protect and enhance the health of the lagoons and to promote education and stewardship. In addition, several large-scale management and restoration efforts, such as the Batiquitos Lagoon Enhancement Project (completed 1996) and the San Dieguito Wetlands Restoration Project (completed 2011), have been undertaken with goals such as the improvement of water quality or enhancement of wildlife habitat within the lagoons.

4. BUENA VISTA LAGOON

Buena Vista Lagoon. (photo by Sean Baumgarten, January 2013)

LOCATED ON THE BOUNDARY BETWEEN THE CITIES OF OCEANSIDE AND CARLSBAD, BUENA VISTA LAGOON EXTENDS NEARLY ONE AND A HALF MILES INLAND AND COVERS APPROXIMATELY 200 ACRES. Several transportation corridors – Interstate 5, the Coast Highway, and the Santa Fe Railroad – cross the lagoon, dividing it into four connected basins. The community of St. Malo sits on a beach berm at the western end, and a weir, originally constructed in the 1940s, separates the lagoon from the ocean.

Today Buena Vista is a freshwater/brackish lagoon, composed of large areas of open water and freshwater marsh. The lagoon is fed by Buena Vista Creek, a perennial, spring-fed stream originating in the San Marcos Mountains east of Vista. With a drainage area of just over 20 square miles, Buena Vista has the smallest watershed of the six lagoons studied.

Timeline: Buena Vista Lagoon

1881-82	California Southern Railroad line constructed between National City and Oceanside.
1883	John Frazier settles on the south shore of Buena Vista Lagoon. His discovery of underground mineral water turns the area into a popular railroad stop known as "Frazier's Station," later renamed Carlsbad.
1901-02	The California Salt Company initiates an ill-fated enterprise to harvest salt from evaporation ponds in Buena Vista Lagoon.
1908	A stand of eucalyptus known as the Hosp Grove, originally intended for commercial harvest, is planted on the southeast side of the lagoon.
1912-15	Pacific Coast Highway constructed.
1914	The Oceanside Mutual Water Company, created in 1914, pumps water from the San Luis Rey Valley to irrigate cropland in Oceanside and Carlsbad.

Urban development around Buena Vista Lagoon. The railroad and highway berms and the remnant salt evaporation ponds are visible in the foreground of this undated photograph **(bottom left)**. The community of St. Malo, built on a berm at the western end of the lagoon, is shown in this photograph from July 27, 1946 **(bottom right)**. (left: photo #267-6, courtesy of Oceanside Historical Society; right: Leeds 1946, courtesy of Special Collections & Archives, UC Riverside)

1928	Sewage effluent from treatment ponds owned by the Vista Sanitation District begins flowing into Buena Vista Creek.
1939	San Diego County outlaws hunting around the lagoon, making the area into a bird sanctuary named after local resident Maxton Brown.
1940s	Weir constructed across the mouth of Buena Vista lagoon, separating it from tidal influence and transforming it into a freshwater/brackish lagoon. The weir is replaced in the 1970s.
1956	The Carlsbad Sanitary District begins discharging sewage effluent into the lagoon. Sewage discharge into the lagoon ceases in 1965.
1965-67	Interstate 5 constructed.
1968	Buena Vista Lagoon becomes the first area added to the California Department of Fish and Game's Ecological Reserve system.
1981	Buena Vista Lagoon Foundation established.

Salt works were developed in the early 20th century at Buena Vista Lagoon. A ca. 1902 photograph **(top left)** shows pumping facilties at a lagoon near Oceanside, likely Buena Vista. A 1950s photograph **(top right)** shows remnant salt evaporation ponds at Buena Vista Lagoon. (left: Bailey 1902; right: courtesy of Oceanside Historical Society)

Sources: Bailey 1902, California Development Board et al. 1923, Storie and Carpenter 1929b, ver Planck 1958, Stanbro 1971, Applegate 1985, Marcus 1989, Ohara 2011, Buena Vista Lagoon Foundation n.d.

EDELIA HOUSE
(CHAUNCEY HAYES)

S O U T H   O C E A N S I D E

B u e n a

d   *V i s t a*   e   *V a l*

a

c

b

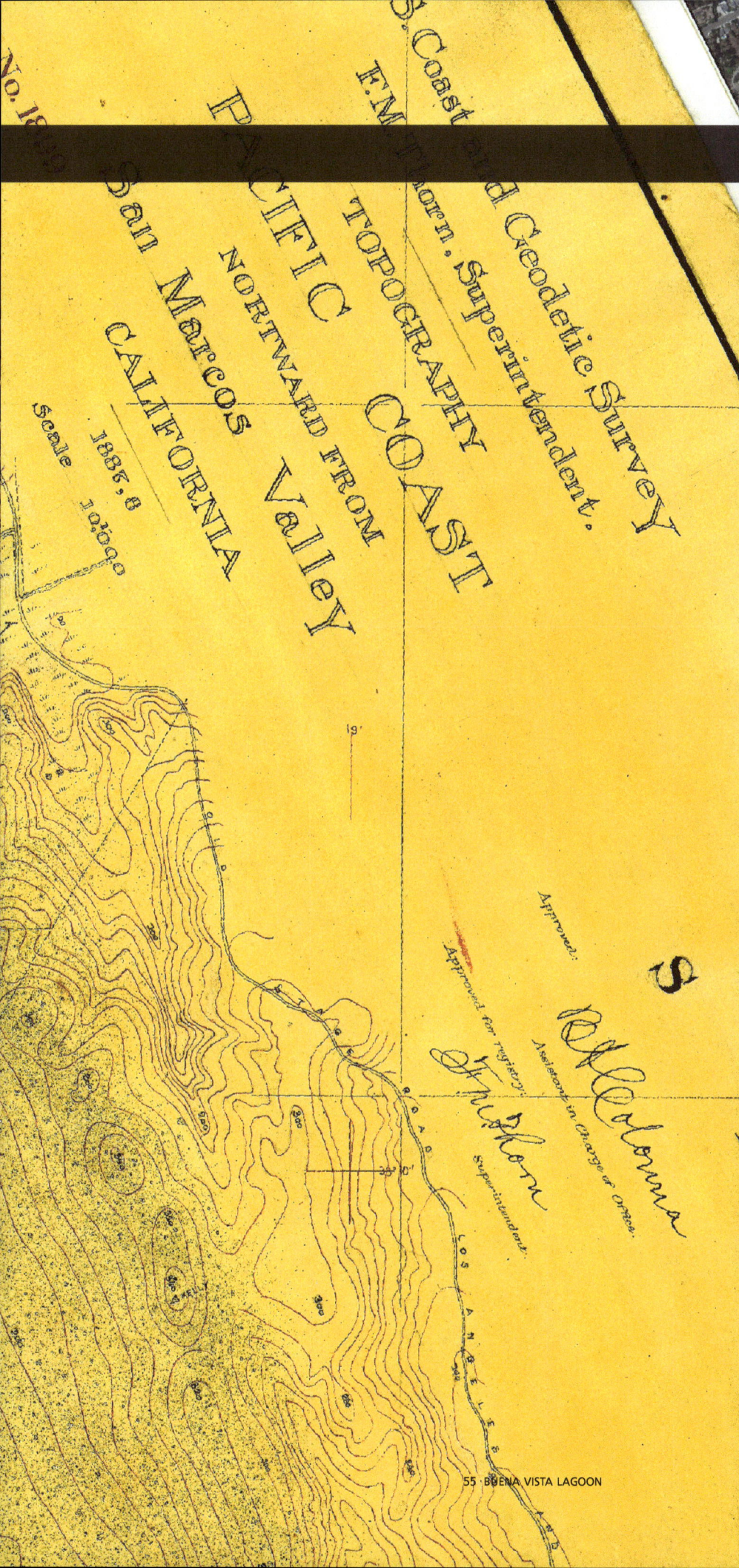

Buena Vista Lagoon as shown on the T-sheet, surveyed August to December 1887 and May to November 1888. (Rodgers and McGrath 1887-8b; courtesy of NOAA)

a The presence of a road along the beach suggests that travel was possible along the coast throughout much of the year, implying that inlet opening may have been relatively infrequent (though crossings could likely have occurred at low tide even if the berm was otherwise submerged). The inlet is shown as closed on the T-sheet.

b The concentric circles shown in channel segments near the coast indicate perennial open water, suggesting that water would have stayed ponded when the lagoon was closed.

c The California Southern Railroad berm cut across much of the lagoon by this time, restricting flow between the lagoon and the ocean. Only a small bridge is shown, apparently over a road through the marsh.

d Numerous roads crossed the salt flat at Buena Vista Lagoon, and even portions of the marsh. Though these roads would likely have been impassable during much of the wet season, they would have provided a speedy route across the lagoon when the salt flat was dry.

e The boundary shown on the T-sheet between the salt marsh fringing the salt flat and upland areas appears to be purposefully imprecise (sketched here in pencil), perhaps indicating a more gradual or variable transition in these areas.

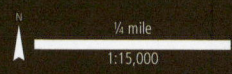

N
¼ mile
1:15,000

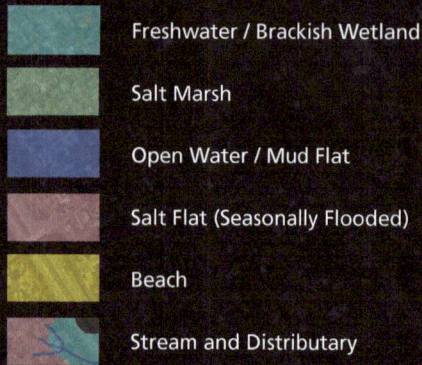

BUENA VISTA LAGOON: Historical Synthesis Overview

Legend:

- Freshwater / Brackish Wetland
- Salt Marsh
- Open Water / Mud Flat
- Salt Flat (Seasonally Flooded)
- Beach
- Stream and Distributary

¼ mile
1:15,000

BUENA VISTA LAGOON HISTORICALLY OCCUPIED ABOUT 310 ACRES, MAKING IT THE SMALLEST OF THE SIX LAGOONS. Its most prominent feature was an extensive salt flat that covered 75% of the lagoon. This salt flat occupied the central and eastern portions of the lagoon and was mostly unvegetated, with small patches or "islands" of marsh supported on higher ground within the flat. Pickleweed-dominated salt marsh was supported in a narrow (<~100 feet) fringe ringing the salt flat, as well as in a larger area (~50 acres) between the salt flat and the beach. In total, salt marsh comprised approximately 23% of Buena Vista Lagoon.

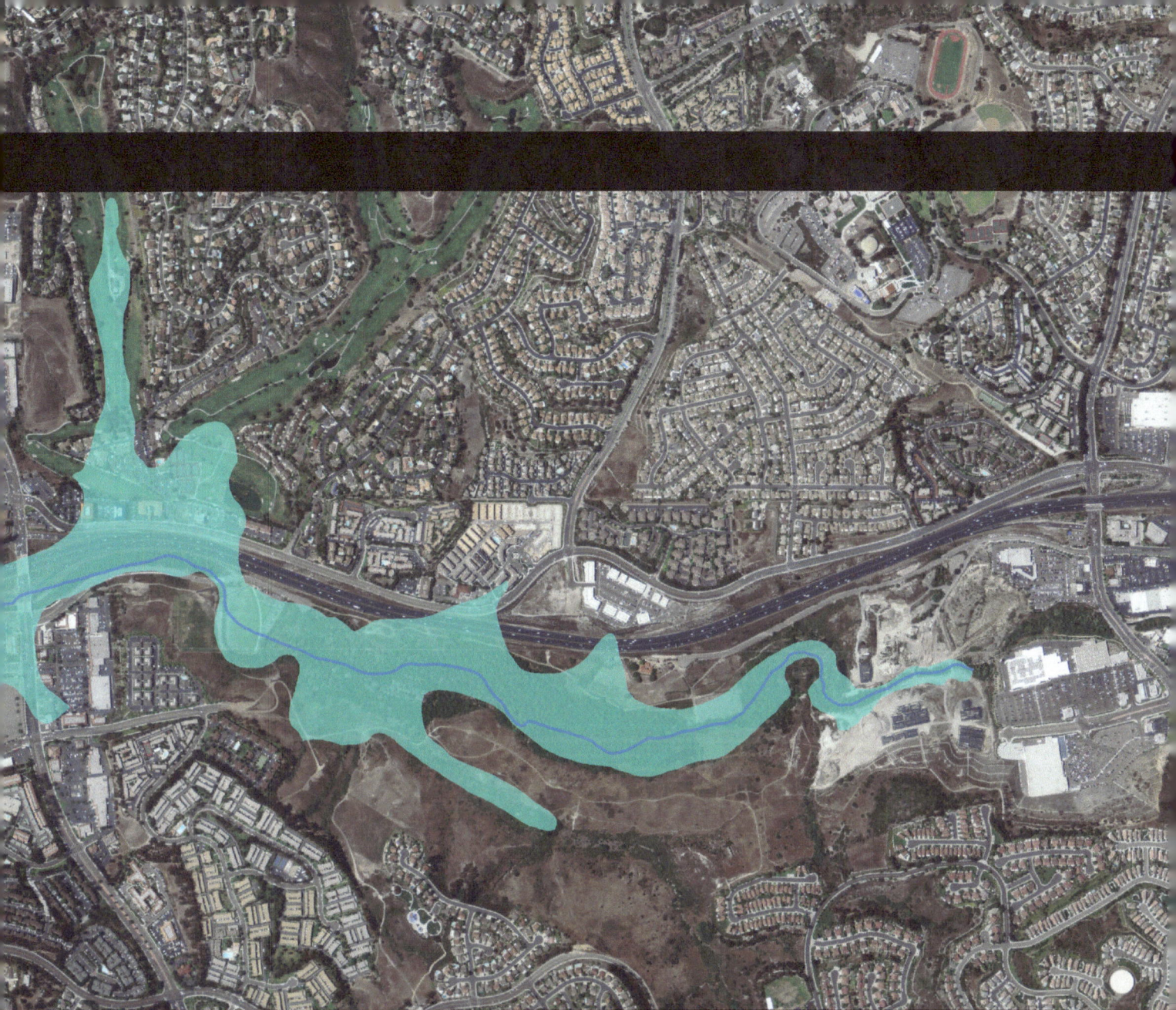

Two distinct channels, shown here as elongate ponds adjoining branches of the salt flat that extended into the marsh plain, meandered through the marsh. When the mouth was open, these features would have acted as inter-tidal channels conveying water between the lagoon and ocean. When the lagoon was separated from the ocean (as shown here), water backed up behind the dunes and ponded in the channels. These channel segments were relatively deep compared to other portions of the lagoon, and thus often maintained perennial open water (Rodgers and McGrath 1887-8b; USGS 1893, [1891]1898). During many summers, these ponds may have constituted the only standing water in the lagoon.

On the eastern edge of the salt flat, an extensive zone of freshwater/brackish transitional wetland – nearly equaling the area of the lagoon itself – extended for almost two miles along Buena Vista Creek. Though the transition from salt marsh to freshwater/brackish wetland at the upland edge of the lagoon looks abrupt on this map, in reality it would have been characterized by a gradual gradient of decreasing salinity. Pickleweed and saltgrass would have been common near the lagoon, with saltgrass likely also common further upstream in "more poorly drained spots" (Storie and Carpenter 1929a,b). Rushes and tules may have occurred where salt concentrations were not too high: Crespí, camped in the Buena Vista Valley with the Portolá Expedition in July 1769 not far from the lagoon, noted "water empounded within a tule-rush patch" (Crespí and Brown 2001; see also Fages and Priestley 1937; Carrico 1977).

SALT FLAT: Salt flat covered 230 acres at Buena Vista Lagoon. Situated at an elevation on average only a little above Mean High Water, much of the flat would have been exposed to regular tidal flooding when the inlet was open; during periods of inlet closure, the flat crystallized salt, forming a "white coating of alkali" (Rodgers 1887-8b). This feature persisted well into the 1920s and 1930s (see page 60).

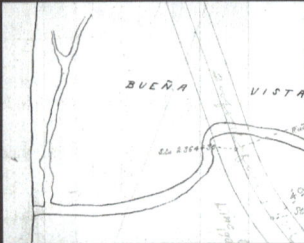

INLET DYNAMICS: Early records document that for much of the year a beach barrier separated Buena Vista Lagoon from the ocean. During the rainy season, however, large flood events periodically breached the sand berm and exposed the lagoon to direct tidal influence (Rodgers 1887-8b, Wood 1913; see page 176).

SALT MARSH FRINGE: Though the main expanse of salt marsh was on the seaward side of the lagoon, a narrow fringe of marsh extended well inland, encircling the salt flat. USCGS Surveyor Rodgers (1887-8b) noted "it is only along the margin nearest the hills, that even the most hardy vegetation is able to maintain itself."

CHANNELS: When the mouth was closed, water ponded behind the beach in these channels, creating elongate perennial ponds within the salt marsh.

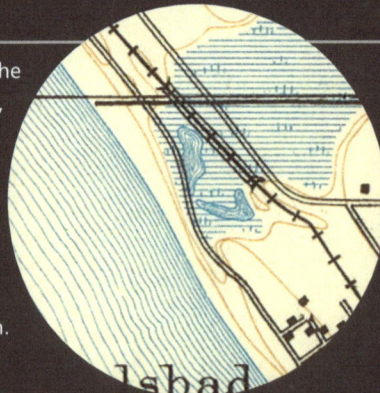

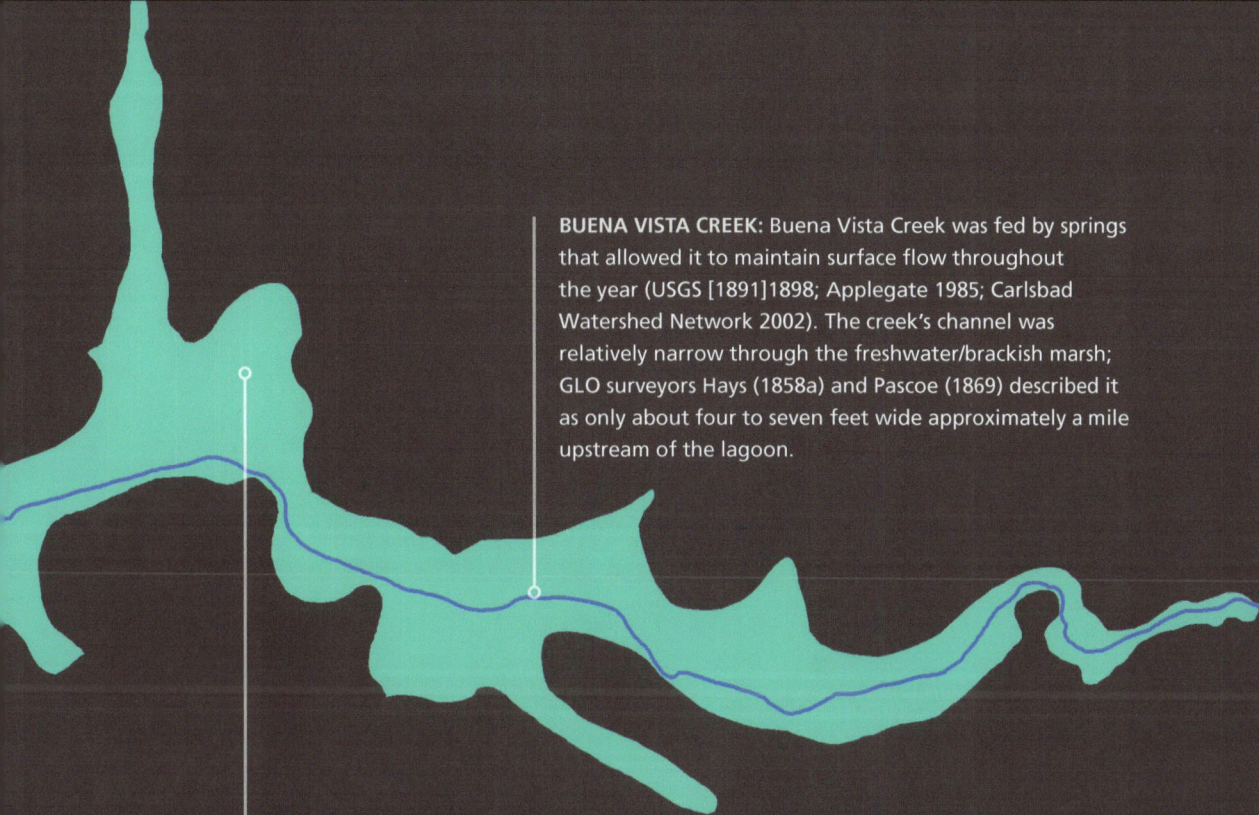

BUENA VISTA CREEK: Buena Vista Creek was fed by springs that allowed it to maintain surface flow throughout the year (USGS [1891]1898; Applegate 1985; Carlsbad Watershed Network 2002). The creek's channel was relatively narrow through the freshwater/brackish marsh; GLO surveyors Hays (1858a) and Pascoe (1869) described it as only about four to seven feet wide approximately a mile upstream of the lagoon.

FRESHWATER/BRACKISH WETLANDS: Early sources indicate that this zone was a mosaic of perennial and seasonal brackish and freshwater wetlands (USGS 1891, Storie and Carpenter 1929a,b). In July 1769, Pedro Fages noted that the valley was "swampy and better supplied with water" than the area above Batiquitos Lagoon, where it was necessary to dig in the dry stream bed to find water (Fages and Priestley 1937).

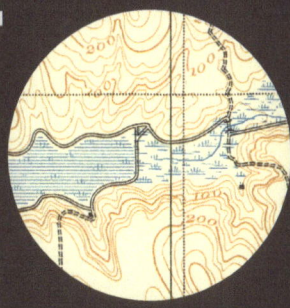

	Freshwater / Brackish Wetland
	Salt Marsh
	Open Water / Mud Flat
	Salt Flat (Seasonally Flooded)
	Beach
	Stream and Distributary

N

¼ mile
1:15,000

Inlet Dynamics: Osgood 1881b, courtesy of California State Railroad Museum; Channels: USGS 1893; photo #HP0673.001, courtesy of Carlsbad City Library Carlsbad History Room; Freshwater/Brackish Wetlands: USGS [1891]1898; Salt Flat: Photo #F_O-2885_2-28-1932, courtesy of The Benjamin and Gladys Thomas Air Photo Archives, Fairchild Collection, UCLA Department of Geography.

BUENA VISTA LAGOON: Salt Flat

An extensive seasonally flooded salt flat dominated Buena Vista Lagoon. The flat was characterized by low-gradient topography, high salinity, and a lack of vegetation. Freshwater inflows often flooded the flat during the rainy season, while evaporation caused the flat to dry out during the summer (Rodgers 1887-8b). In addition to these seasonal fluctuations, the extent and duration of flooding also varied annually.

The salt flat was first documented in the 1850s by GLO Surveyor Freeman, who described a "flat" with "no vegetation" which was "subject to inundations by high tides" (Freeman 1854a). Rodgers (1887-8b) subsequently described the salt flat in detail while completing the T-sheet survey, noting the presence of a "white coating of alkali, which is probably a residium of the moisture evaporated during the hot sunny days of summer." He went on to compare the salt flat to the "'Alkali Plains' of Nevada and Utah." This similarity was also noted by Carpelan (1969) nearly a century later, who wrote, "This sort of lagoon when evaporated to dryness becomes a salt pan as flat as a desert dry lake."

The salt flat persisted as Buena Vista Lagoon's dominant feature well into the 20th century. At the time of this photograph, taken in February 1932, the flat appears to have been shallowly flooded. Open water channels can be seen threading through the salt marsh and passing under the railroad and the Coast Highway on the left side of the image.

Photo #F_O-2885_2-28-1932, courtesy of The Benjamin and Gladys Thomas Air Photo Archives, Fairchild Collection, UCLA Department of Geography.

61 BUENA VISTA LAGOON

Local residents recalled the salt flat drying out seasonally or during dry years (Hinman 2012; Howard-Jones 1982). Allan Kelly, whose family owned the nearby Agua Hedionda Rancho in the early 1900s (Anderson 2007), described people racing cars across the bottom of Buena Vista Lagoon in the summer:

> I can well remember the 'race track' in the Buena Vista lagoon…In the summer as soon as the floor of the lagoon was dry enough, they began driving straight across the bottom to the foot of the hill by the old graveyard. This was the only straight, smooth road in the north end of the county and the only place where the sports could get up 'full speed' in their new-fangled Stanley Steamers, Thomas Flyers, Merry Oldsmobiles, Pope Hartfords and such. (Kelly 1959; see also Harmon 1967)

Attempts were made to harvest salt from the flat for commercial use. In 1901 the California Salt Company leased most of Buena Vista Lagoon for this purpose:

> It is the intention of the company to at once commence the laying out of immense vats in the slough for making salt from sea water, and, if the business warrants, to erect a large refinery at the place for the purpose of handling the salt. (*San Francisco Call* 2/12/1901)

Vats, wells, and pumps were constructed, but leakage from the evaporation ponds proved problematic, and the scheme ultimately failed (Bailey 1902, ver Plank 1958).

(left) Buena Vista Lagoon is shown as two separate water bodies in this 1872 map, perhaps representing open water features in the marsh near the coast and a separate, disconnected open water "lagoon" occupying the salt flat further inland. (Wheeler et al. 1872, courtesy of San Diego Cartographic Services)

(far left) Children playing on the banks of Buena Vista Lagoon. The unvegetated salt flat is visible in the background of this ca. 1920s photograph, along with a fringe of marsh vegetation. The well and ponded water in the foreground may suggest the presence of high groundwater. (Photo # HP1499.012, courtesy of Carlsbad City Library Carlsbad History Room)

Buena Vista Creek has formed a wide valley, the upper end with a lagoon, the lower with marshland. During the summer the creek from the lagoon is too meager to connect with the marshland, but in winter the surrounding areas drain into the creek which overflows and the entire marshland may be under several feet of water…The marsh is cut off by low-lying sand dunes from the ocean except at very high tides…The creek channel lies in the center of the valley with the lower stretches of marsh covered by water in winter. In summer the channel only is partly filled with water, the rest of the marshland consisting of large, flat salt-covered areas without vegetation.

—PURER 1942

SINCE THE 19TH CENTURY, THE MOSAIC OF HABITAT TYPES COMPOSING BUENA VISTA LAGOON HAS BEEN ALTERED CONSIDERABLY. The most significant changes were set in motion in the 1940s, when a weir was constructed at the mouth of the lagoon, permanently separating Buena Vista from the ocean and allowing water in the lagoon to be maintained at a relatively constant level throughout the year (Cain 1982, Marcus 1989). Installation of the weir drastically transformed lagoon habitats, virtually eliminating salt marsh and extensive salt flats and replacing them with open water (ranging from fresh to brackish) and freshwater/brackish marsh dominated by cattails (*Typha latifolia*) and California bulrush (*Schoenoplectus californicus*; Everest International Consultants, Inc. 2004), even as most of the freshwater/brackish wetlands that historically existed upslope of the lagoon were eliminated due to development. Today, sedimentation is contributing to the expansion of cattails and bulrush into open water portions of the lagoon.

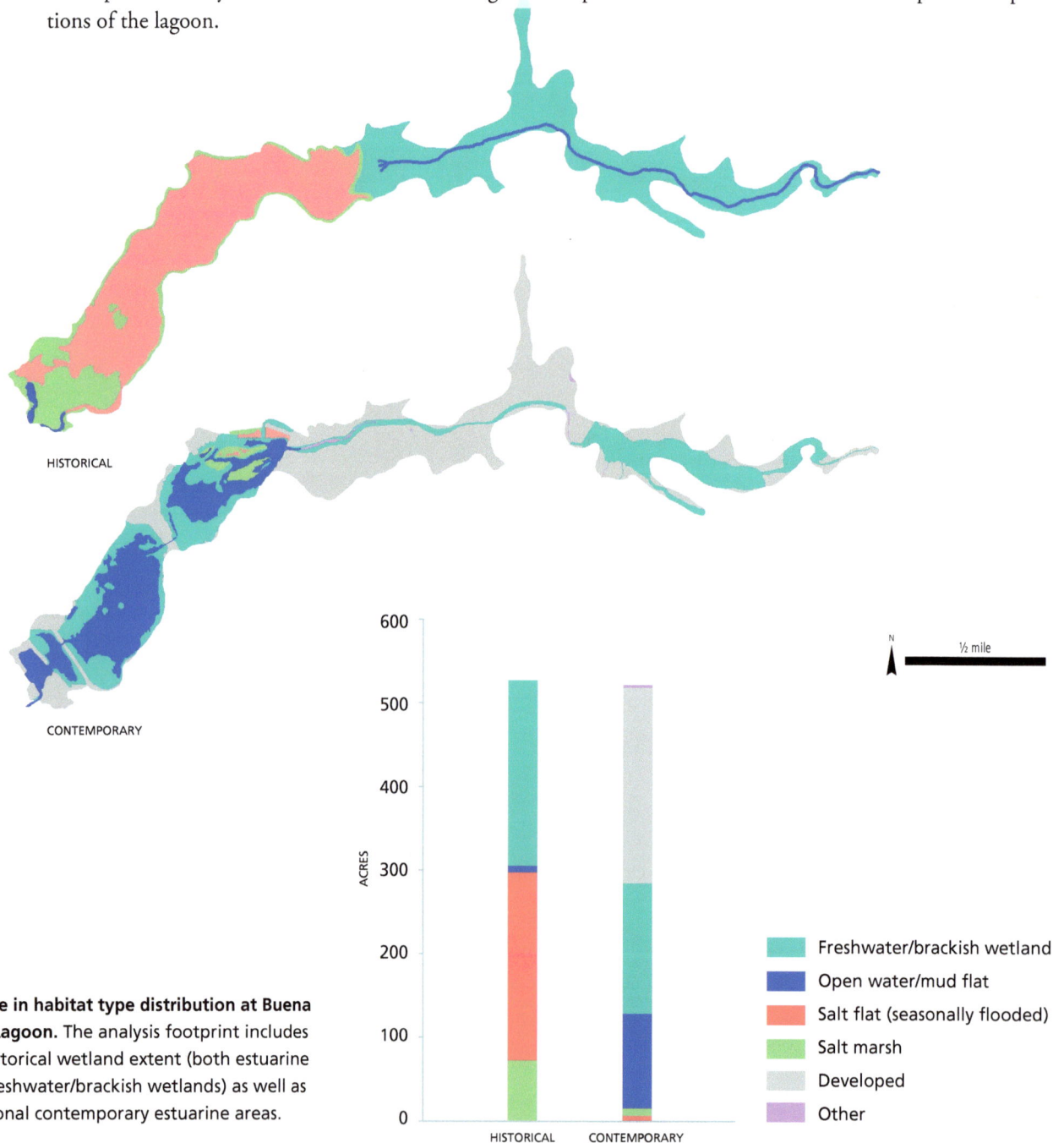

HISTORICAL

CONTEMPORARY

Change in habitat type distribution at Buena Vista Lagoon. The analysis footprint includes the historical wetland extent (both estuarine and freshwater/brackish wetlands) as well as additional contemporary estuarine areas.

Legend:
- Freshwater/brackish wetland
- Open water/mud flat
- Salt flat (seasonally flooded)
- Salt marsh
- Developed
- Other

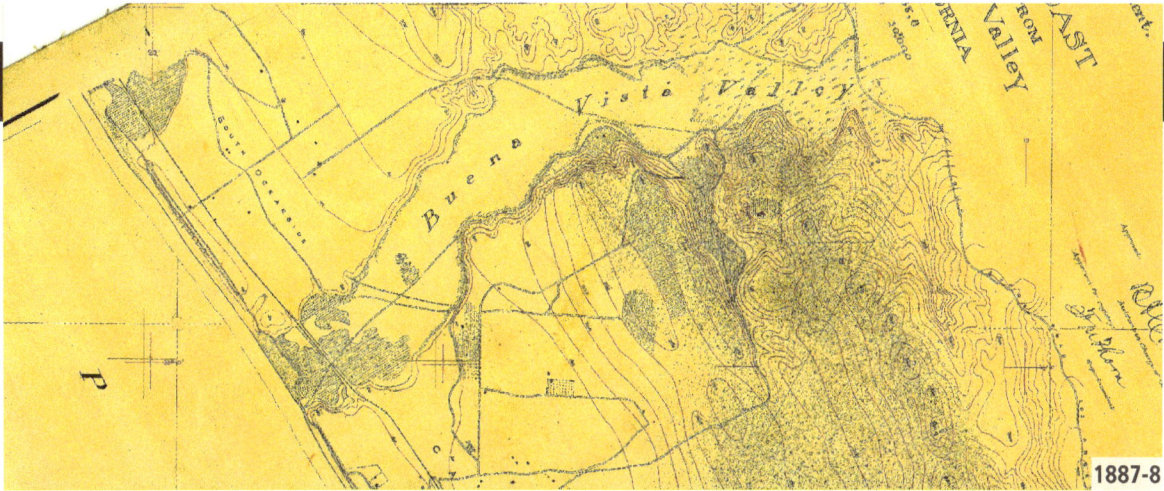

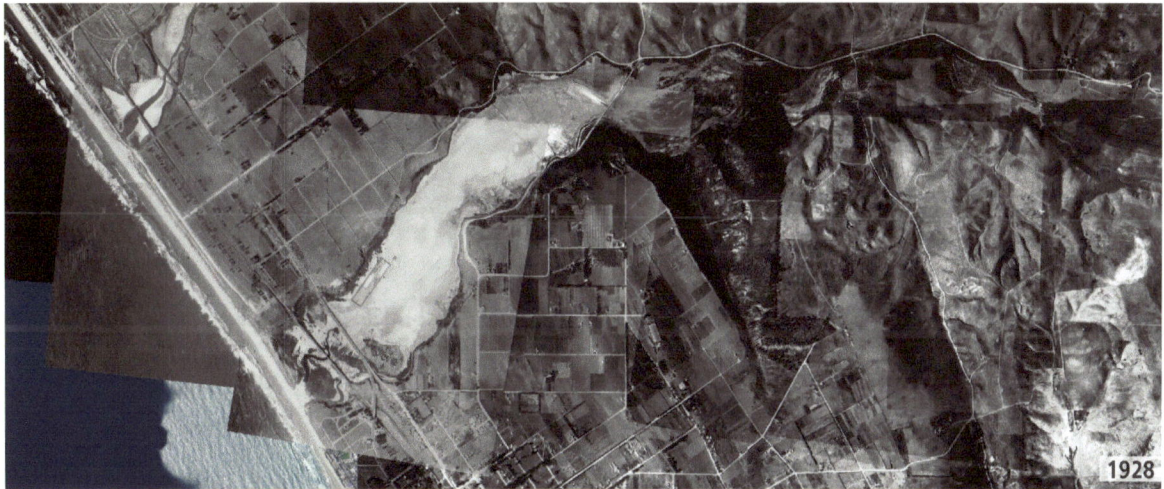

Top: In 1881-2, the California Southern Railroad was built across Buena Vista Lagoon, restricting lagoon circulation. The 1887-8 T-sheet shows that a single short bridge provided the only pathway for fluvial and tidal flows. **Middle:** An extensive salt flat was still present in 1928, when these aerial photos were taken. The dynamic transition zone between the estuarine habitats of the lagoon and the fluvial/terrestrial habitats further inland is visible at the eastern edge of the salt flat. **Bottom:** Though sewage discharge to the lagoon stopped in 1965 (Stanbro 1971), increased discharge from irrigation runoff and from Buena Vista Creek has maintained perennial water in the non-tidal lagoon. (Rodgers and McGrath 1887-8b, San Diego County 1928, NAIP 2009)

5. AGUA HEDIONDA LAGOON

The name *Agua Hedionda* – literally "stinking water" in Spanish – has been used to infer that the lagoon was foul-smelling in the past, though the name in fact refers to a nearby sulfur springs (Capace 1999, Fetzer 2005). The Portolá Expedition named the valley and the lagoon El Beato Simón de Lipnica ("Blessed Simon of Lipnica") (Crespí and Brown 2001, Fetzner 2005).

Agua Hedionda Lagoon. (photo by Sean Baumgarten, January 2013)

AGUA HEDIONDA LAGOON OCCUPIES 400 ACRES JUST SOUTH OF DOWNTOWN CARLSBAD, AND EXTENDS INLAND APPROXIMATELY 1.7 MILES. It is primarily fed by Agua Hedionda Creek, and has the second-smallest drainage area (approximately 30 square miles) of the six lagoons. Interstate 5, the Coast Highway, and the Santa Fe Railroad cross the lagoon, dividing it into three connected basins. Over 75% of the contemporary estuarine area is open water, with the remainder comprised of marsh and mud flat.

Jetties and regular dredging keep the mouth of the lagoon permanently open to the ocean. The Encina Power Station, which sits on the lagoon's south shore, pumps cooling water from the lagoon. A desalination plant currently under construction will also obtain water from the lagoon.

Timeline: Agua Hedionda Lagoon

1842	Rancho Agua Hedionda granted to Jose María Romualdo Marrón. The Rancho extends approximately 5.5 miles inland and encompasses about 21 square miles, including all of Agua Hedionda Lagoon.
1870	Robert Kelly, a veteran of the Mexican-American War, acquires Rancho Agua Hedionda. Gunn (1887) reports that the ranch is "devoted to stock-raising."
1881-82	California Southern Railroad line constructed between National City and Oceanside.
1890s	Rancho Agua Hedionda divided among Kelly's heirs.
1912-15	Pacific Coast Highway constructed.
1914	The Oceanside Mutual Water Company, created in 1914, pumps water from the San Luis Rey Valley to irrigate cropland in Oceanside and Carlsbad.
1915	A concrete bridge constructed across the mouth of Agua Hedionda Lagoon, causing sediment to build up in the channel.
Early 1930s	New highway bridge constructed; old bridge dynamited into channel.
1954	Encina Power Station built on south shore of Agua Hedionda Lagoon.
1954	Agua Hedionda lagoon dredged and jetties constructed to keep the mouth permanently open to the ocean. Water from the lagoon used to provide cooling water for the Encina Power Station.
1965-67	Interstate 5 constructed.
1990	Agua Hedionda Lagoon Foundation established.

Sources: California Development Board et al. 1923, Storie and Carpenter 1929b, Kelly 1959, Ritter 1963, Kelly in Harmon 1967, Howard-Jones 1982, Marcus 1989, Hawthorne 2003, Christenson and Sweet 2008

1955 photograph showing newly constructed Encina Power Station (bottom left) and recently dredged lagoon. Buena Vista Lagoon can be seen in the background. (photo #UT 8248-254, courtesy of San Diego History Center)

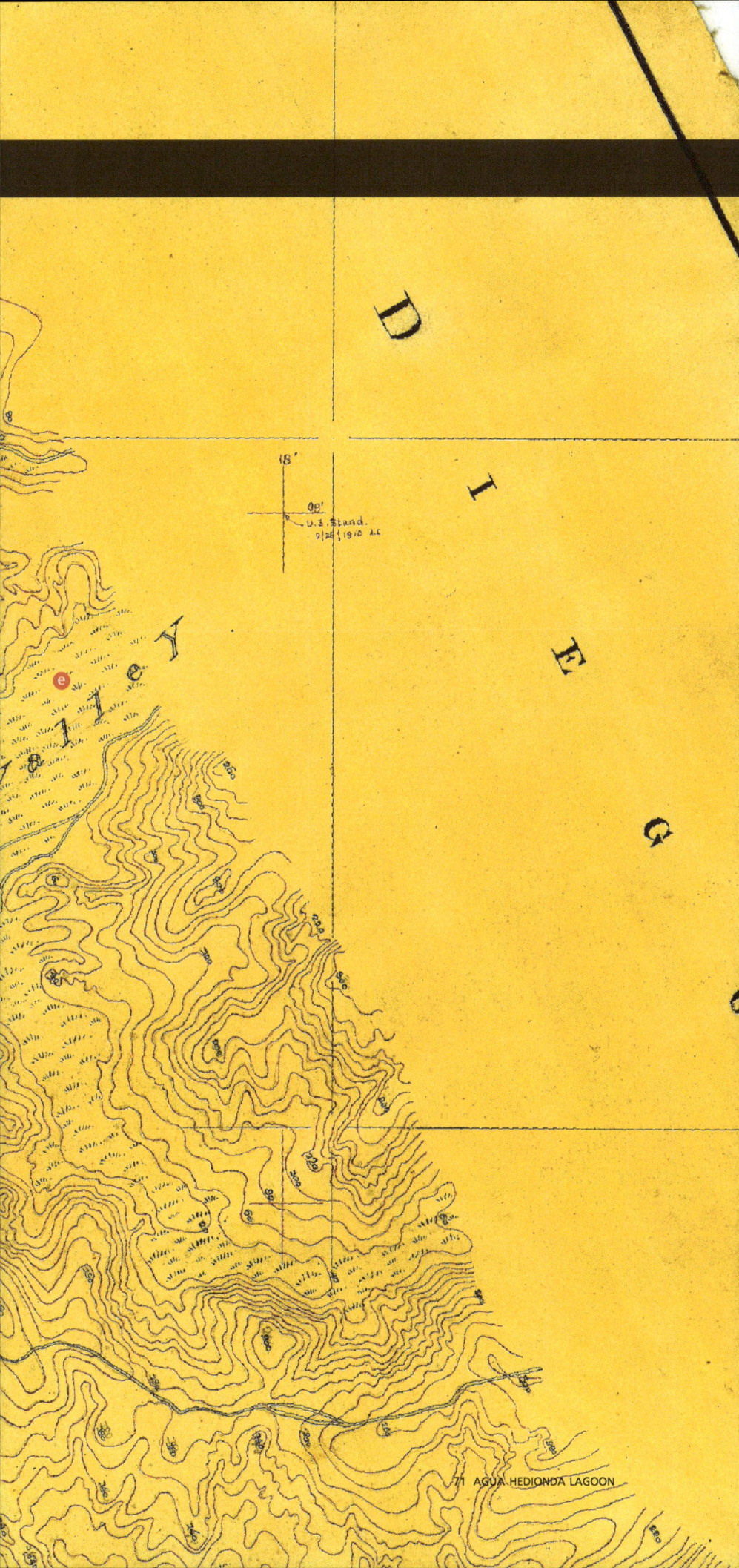

Agua Hedionda Lagoon as shown on the T-sheet, surveyed August to December 1887 and May to November 1888. (Rodgers and McGrath 1887-8b; courtesy of NOAA)

a The T-sheet shows the inlet as closed, with a road running along the beach between the lagoon and the ocean.

b Some of the ponds in the northwestern part of the lagoon are shown with the perennial water symbol, indicating that these areas provided open water habitat year-round.

c While the northern channel connecting the salt flat to the ocean has been bisected by the railroad berm, a bridge is shown over the southern channel.

d As with Buena Vista Lagoon, the line between the salt flat and the fringing salt marsh appears to be intentionally indistinct, suggesting a gradual transition.

e The valley floor on the upland side of the lagoon is depicted using the symbol for herbaceous vegetation. Other early sources indicate that this area constituted the lowermost portion of an extensive freshwater/brackish transitional wetland.

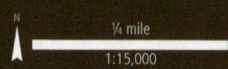

D I E G O

Valley

18'

90'
U.S. Stand.
9/26/1910 A.C

N

¼ mile
1:15,000

71 AGUA HEDIONDA LAGOON

OCCUPYING ABOUT 320 ACRES, AGUA HEDIONDA LAGOON HISTORICALLY CONSISTED OF APPROXIMATELY 51% SEA-
SONALLY FLOODED SALT FLAT, 42% SALT MARSH, AND 7% OPEN WATER/MUD FLAT. The eastern portion of the lagoon
was characterized by a large central salt flat, which flooded seasonally but dried out during the summer months.
Observers viewing the lagoon during the dry season were struck by the unusual appearance of the salt flat: Friar
Juan Crespí, for example, noted a "great deal of white glitter" as he passed by the lagoon in July 1769 (Crespí and
Brown 2001). A century later, USCGS Surveyor Augustus Rodgers described the flat as "white and glistening as
snow" (Rodgers 1887-8b).

While the salt flat was the dominant feature in the eastern part of the lagoon, the western portion of the lagoon
supported large areas of salt marsh. Several open water channels, varying in width from approximately 40-180 feet,

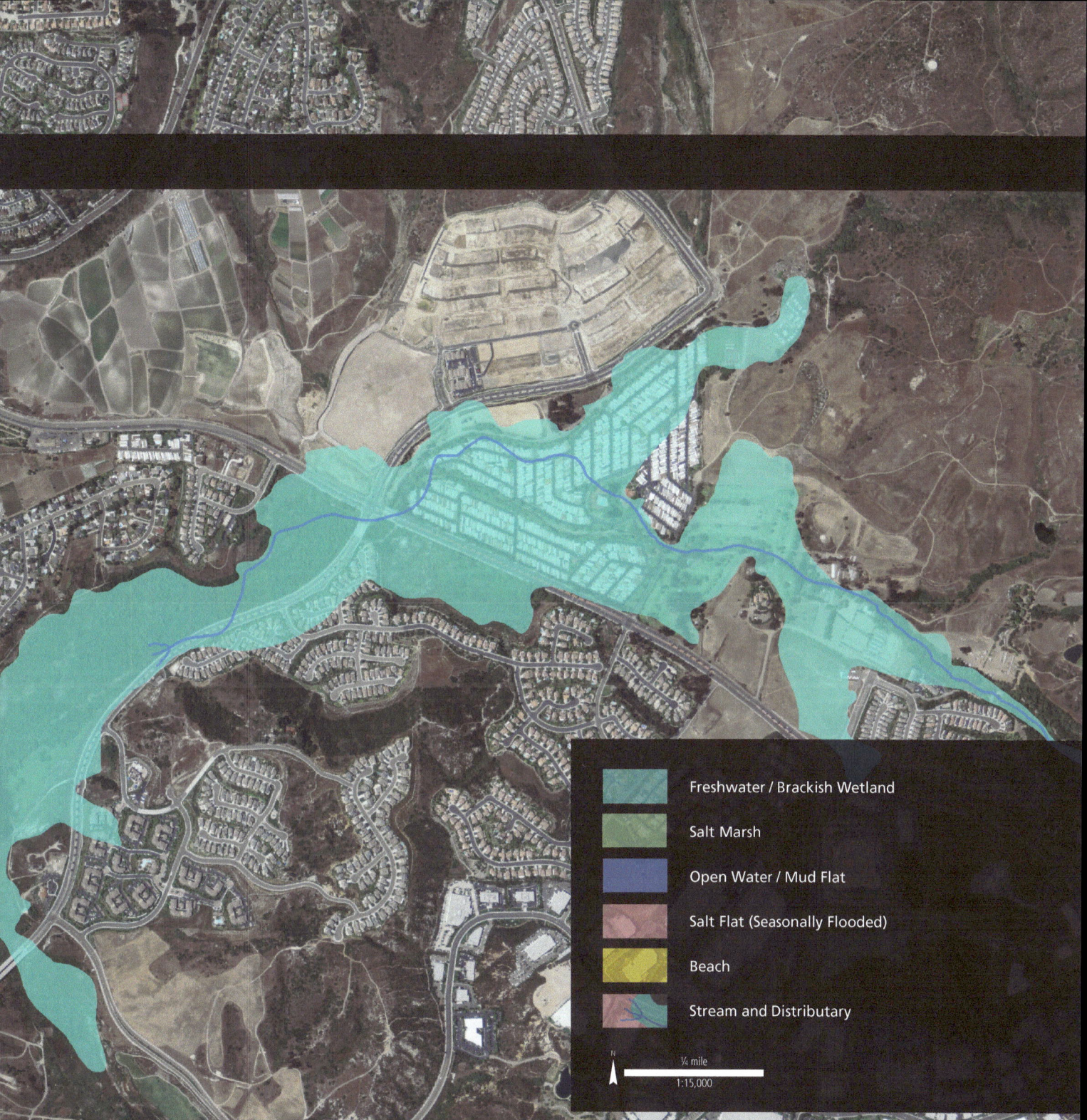

	Freshwater / Brackish Wetland
	Salt Marsh
	Open Water / Mud Flat
	Salt Flat (Seasonally Flooded)
	Beach
	Stream and Distributary

¼ mile
1:15,000

wound through the western salt marsh. These channels likely retained water throughout much of the year, though during dry periods large portions of the channels would have become mud flats. In the northwestern corner of the lagoon a series of open water ponds marked the path of a channel that at times would have connected the lagoon to the sea.

An extensive freshwater/brackish transitional wetland adjoined the lagoon on the eastern side, and continued for over two and a half miles up the valley along Agua Hedionda Creek. The transitional wetland was underlain by several different types of alkali-influenced soils. Areas nearest the lagoon were likely dominated by pickleweed and saltgrass (Storie and Carpenter 1929a,b), while areas further up the valley likely supported less salt-tolerant vegetation.

AGUA HEDIONDA LAGOON: Points of Interest

PERSISTENT PONDS: A series of scoured ponds in the northwestern portion of the lagoon are likely detached channel segments representing a connection to a northern inlet. The northern inlet is visible in sources from the late 1920s/early 1930s, and corresponds to the location where jetties constrain the inlet today. The position of many of these ponds is consistent across more than fifty years of historical mapping.

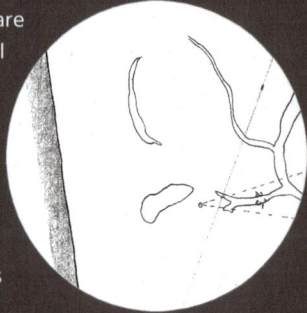

SALT FLAT: One of the first written accounts of Agua Hedionda Lagoon, a journal entry by Friar Juan Crespí from July 1769, describes the flat as a "salt deposit" marked by "white glitter" (Crespí and Brown 2001). The salt flat was described in more detail by surveyor Augustus Rodgers in 1888, who wrote of an alkali plain "as white and glistening as snow" (Rodgers 1887-8b); Rodgers also noted that the elevation of the flat was only slightly above Mean High Water. When the lagoon was parceled out amongst the Kelly family in the 1890s, the "dry salt flats" of the lagoon remained as common property (Kelly 1959). The flat persisted well into the 1930s and 1940s, desiccating seasonally during dry years (Hinman 2012) – as in the 19th century, residents would drive across the dry flat to get from one place to another.

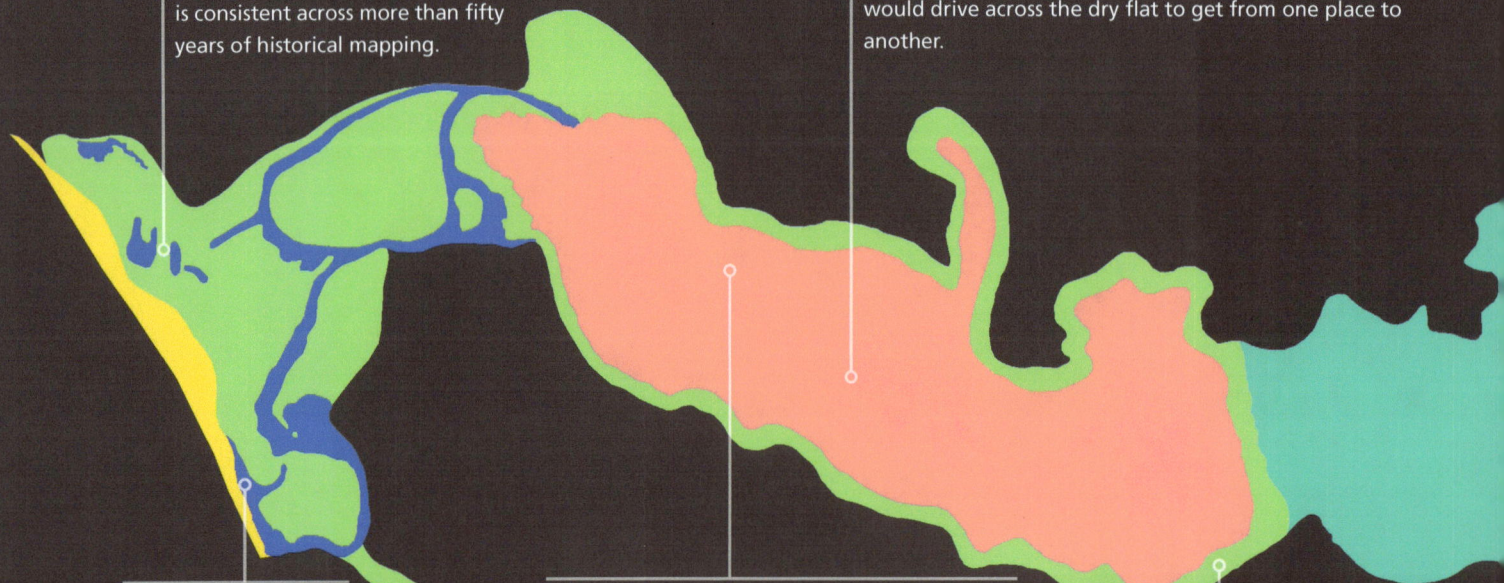

INLET DYNAMICS: Early sources indicate that Agua Hedionda Lagoon was intermittently connected to the ocean through this southern inlet (see page 176).

LOW WATER CHANNEL: A subtle network of shallow channels conveyed water onto and off of the salt flat, as shown on this T-sheet resurvey (top) and 1928 aerial image (bottom).

SALT MARSH FRINGE: Salt marsh vegetation was found in a narrow band fringing the lagoon. As with Buena Vista to the north, surveyor Rodgers noted, "It is only along the margin nearest the hills, that even the most hardy vegetation is able to maintain itself" (Rodgers 1887-8b).

Persistent Ponds: Osgood 1881a, courtesy of California State Railroad Museum; Inlet Dynamics: Unknown 1881, courtesy of California State Railroad Museum; Low Water Channel top: Knox 1934a; Low Water Channel bottom: San Diego County 1928; Agua Hedionda Creek top: USDC ca. 1840b, courtesy of The Bancroft Library; Agua Hedionda Creek bottom: HP0681.001, courtesy of Carlsbad City Library Carlsbad History Room.

Legend

- **Freshwater / Brackish Wetland**
- **Salt Marsh**
- **Open Water / Mud Flat**
- **Salt Flat (Seasonally Flooded)**
- **Beach**
- **Stream and Distributary**

N
¼ mile
1:15,000

AGUA HEDIONDA CREEK: Portions of Agua Hedionda Creek may have maintained surface flow or standing water year-round (USGS [1891] 1898), though continuous surface flow probably ceased in many reaches during the summer months. In July 1769, Friar Juan Crespí reported that in the valley above Agua Hedionda Lagoon there was "no running water" in Agua Hedionda Creek, "only a good amount standing emponded" (Crespí and Brown 2001). Passing through in April of the following year, however, Crespí observed "a good-sized stream of water at the Beato Simón hollow" (*El Beato Simón de Lipnica* was the name given to Agua Hedionda valley by the Portolá Expedition). He also noted the presence of "a great deal of sycamore trees here in the stream hollow" (Crespí and Brown 2001).

In this 1840s diseño (right, top), the thin line depicting Agua Hedionda Creek is shown terminating east of El Camino Real. Riparian trees can be seen flanking the creek, and an *aguaje* (spring) is labeled further up the valley.

An undated photograph (right, bottom) shows flood flows in Agua Hedionda Creek. Several bare sycamore trees line the creek. The photograph was taken along Sunnycreek Road, which parallels Agua Hedionda Creek for a short distance several miles upstream of the lagoon.

TRANSITIONAL WETLANDS: East of the lagoon, the salt flat and fringing salt marsh graded into an extensive freshwater/ brackish transitional wetland (USGS [1891] 1898, 1901b).

Breaching the Agua Hedionda Lagoon inlet. This series of photos from 1950 shows a channel being dug through the beach barrier to establish a connection between the lagoon and the ocean. The photo in the upper left shows the beach barrier prior to breaching, and the bottom photos show the inlet after a new channel has been excavated.

Historically, the mouth of the lagoon likely remained closed for extended periods during the dry season, though early sources reveal a pattern of seasonal opening. For example, Surveyor Augustus Rodgers, writing of Buena Vista, Agua Hedionda, and Batiquitos lagoons, stated, "They are … protected now [summer] from the break of sea waves by dykes of sand or shingle. During the wet season, they are overflowed by fresh water and storm waves break over the front dykes mentioned." (Rodgers 1887-8b; photos #HP1550.024 through HP1550.027, courtesy of Carlsbad City Library Carlsbad History Room)

CONSTRUCTION OF THE RAILROAD IN 1881-82 AND THE COAST HIGHWAY IN 1912-15 WERE THE FIRST MAJOR DIRECT MODIFICATIONS TO AGUA HEDIONDA LAGOON'S HYDROLOGY AND ECOLOGY (Hawthorne 2003). The berms for these transportation corridors bisected the lagoon, constricting water flow and sediment transport. Interstate 5 was constructed in 1965, further partitioning the lagoon.

Evidence for direct manipulation of the lagoon inlet dates back to at least 1950 (see photographs on previous page). In 1954, jetties were installed to keep the lagoon mouth permanently open to the ocean, and much of the lagoon basin was dredged; these manipulations were undertaken so that the lagoon could be used to provide cooling water for the Encina Power Station.

As a result of dredging, bridge construction, and other impacts, the habitats that make up Agua Hedionda Lagoon today bear little resemblance to the historical suite of habitats. Open water/mud flat constitutes approximately 75% of the lagoon (excluding freshwater/brackish transitional wetlands), with marsh comprising approximately 20%. Salt flat, the dominant habitat type historically, today occupies less than 5% of the lagoon area. The extensive freshwater/brackish wetlands that existed upstream of the lagoon have also been significantly reduced, from over 400 acres historically to approximately 100 acres today.

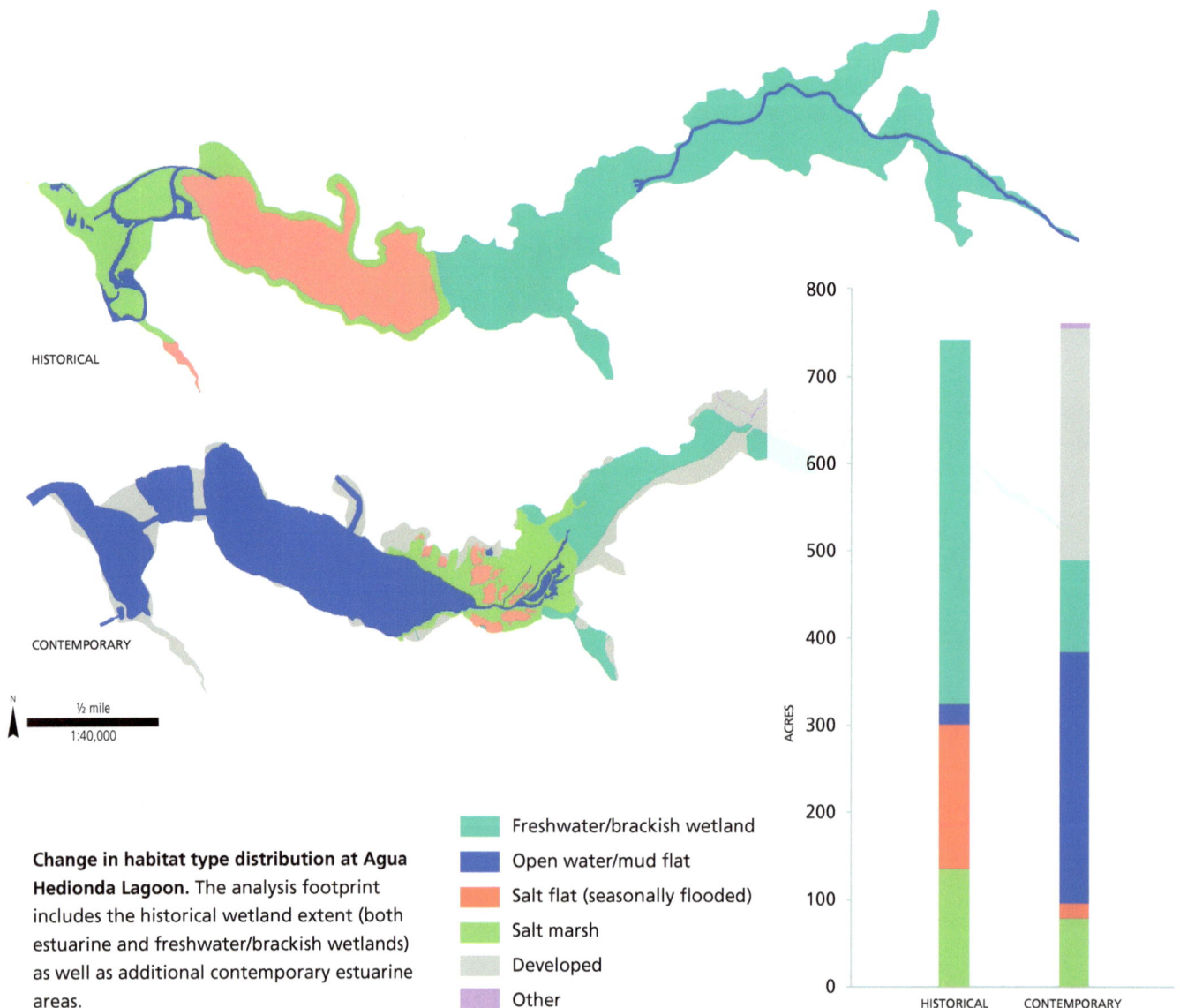

HISTORICAL

CONTEMPORARY

½ mile
1:40,000

Change in habitat type distribution at Agua Hedionda Lagoon. The analysis footprint includes the historical wetland extent (both estuarine and freshwater/brackish wetlands) as well as additional contemporary estuarine areas.

Freshwater/brackish wetland
Open water/mud flat
Salt flat (seasonally flooded)
Salt marsh
Developed
Other

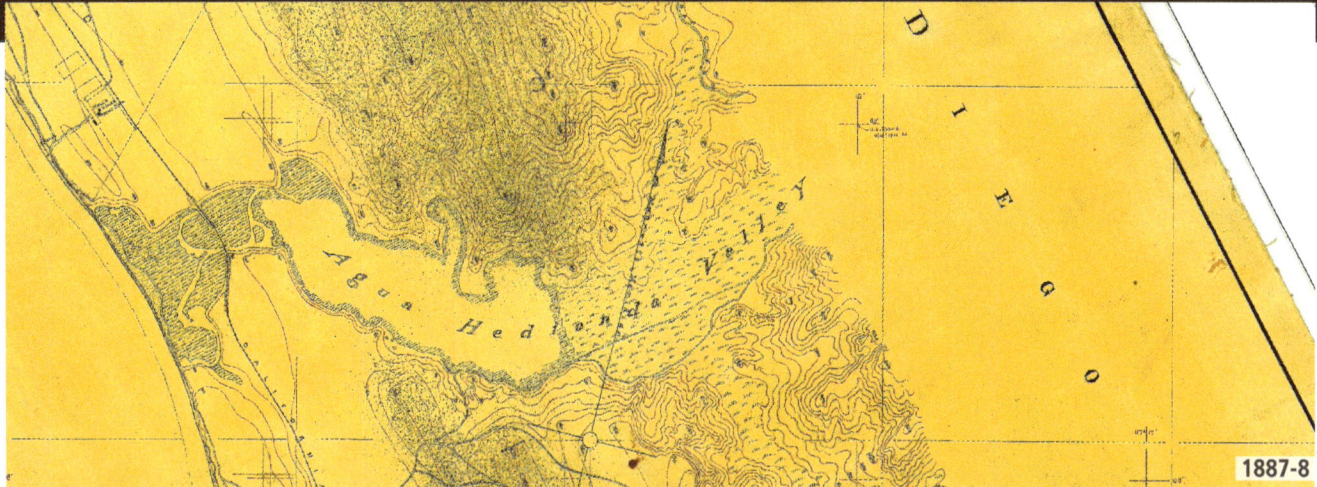

The 1887-8 T-sheet **(top)** shows the lagoon dominated by seasonally flooded salt flat to the east and by salt marsh and open water channels to the west. Much of the historical channel configuration is still visible in the 1928 aerial photos **(middle)**, though the inlet location had been confined to the northern end of the lagoon by Highway 101. Dredging and the construction of jetties in 1954 transformed the lagoon habitats and hydrology. Today the lagoon is characterized by three large open water basins, as shown in the 2009 aerial photos **(bottom)**. (Rodgers and McGrath 1887-8b, San Diego County 1928, NAIP 2009)

6. BATIQUITOS LAGOON

The name *Batiquitos* is derived from *batequi*, used in northwestern Mexico to describe a "hole dug in a riverbed to find water" (Gudde and Bright 1998).

Batiquitos Lagoon covers 600 acres between Carlsbad and Encinitas, extending inland approximately 2.5 miles to El Camino Real. Its watershed covers about 50 square miles – the third smallest of the six lagoons studied – and it receives freshwater input from San Marcos and Encinitas creeks. The Coast Highway and the Santa Fe Railroad cross Batiquitos Lagoon near the coast, while Interstate 5 crosses further inland. La Costa Avenue runs along the base of the bluffs on the south side of the lagoon.

The lagoon was dredged in the mid-1990s as part of an enhancement project, and as a result open water now comprises the majority of the lagoon area.

Batiquitos Lagoon. (photo by Sean Baumgarten, January 2013)

Timeline: Batiquitos Lagoon

1840	Rancho Los Vallecitos de San Marcos, located in Batiquitos Lagoon's watershed approximately seven miles inland, is granted to José María Alvarado.
1875	Nathan Eaton settles the area to the south of Batiquitos Lagoon and begins farming. Known as Eatonville for a time, the town is later renamed Encinitas.
1881-82	California Southern Railroad line constructed between National City and Oceanside.
1901-02	The California Salt Company begins producing salt from 25 acres of evaporation ponds at Batiquitos Lagoon.
1912-15	Pacific Coast Highway constructed.
ca. 1914	Commercial cultivation of vegetable and grains begins in Carlsbad, followed several years later by flowers and avocados.
1952	San Marcos Dam constructed.
1965-67	Interstate 5 constructed.
1960s-70s	Treated wastewater released into the lagoon.
1970s	Recreational vehicle use and helicopter flight testing occur on the salt flat.
1983	Batiquitos Lagoon Foundation established.
1996	Batiquitos Lagoon Enhancement Project completed.

Sources: Bailey 1902, California Development Board et al. 1923, Storie and Carpenter 1929b, Heilbron 1936, Unknown 1975, Mudie et al. 1976, County of San Diego 1979, County of San Diego 1996, Fetzer 2005, Merkel & Associates 2009

By the 1920s, agricultural land uses dominated the coastal areas south of Batiquitos Lagoon, as shown in this ca. 1925 photo. (photo #90:18138-664, courtesy of San Diego History Center)

I have personal experience of the excellence of grapes, figs, oranges, lemons, apples, pears, and peaches, grown entirely without irrigation upon the plateaus between San Alejo and Cottonwood and San Marcos valleys.

—RODGERS 1887-8A

Batiquitos Lagoon as shown on the T-sheet, surveyed August to December 1887 and May to November 1888 (Rodgers and McGrath 1887-8a,b; courtesy of NOAA). Batiquitos Lagoon sits at the boundaries of two T-sheets, which have been stitched together digitally here.

(a) As with its neighbors to the north, Batiquitos Lagoon is depicted with a closed inlet separated from the ocean by a beach berm.

(b) The channel shown in the southwest portion of the lagoon has clearly been diverted by the construction of the railroad berm by the time of the T-sheet survey. Connected to this channel is a network of narrow salt flat "fingers" surrounded by salt marsh.

(c) A road, likely impassable during the wet season, is shown traversing the salt flat.

(d) Unlike the northern two lagoons, no fringing salt marsh is shown around the edge of Batiquitos Lagoon's salt flat. It is not clear if this was a real difference between lagoons, or simply an inconsistency in the mapping.

The greater part of this area is covered by a white coating of alkali... Except where this coating is marked by wagon tracks, it is as white and glistening as snow.

—RODGERS 1887-8B

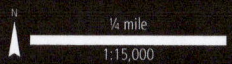

Map labels:

Valley

Register No. 1898

Survey by
Aug:F:Rodgers,Chief of Party
and
John E. McGrath, Sub Assistant

San Diego
CALIFORNIA
1887-8
Scale 10,000

33°05'

¼ mile
1:15,000
N

AN ELONGATE LAGOON OCCUPYING ABOUT 570 ACRES, BATIQUITOS LAGOON WAS HISTORICALLY COM-
PRISED OF APPROXIMATELY 80 ACRES OF SALT MARSH, 480 ACRES OF SALT FLAT, AND 10 ACRES OF
OPEN WATER/MUD FLAT. Its most notable feature was an extensive, seasonally flooded salt flat, which
represented over 80% of the estuarine area – a larger proportion than for any other lagoon in the
study. This salt flat spanned more than two miles from east to west, beginning between the present-
day railroad tracks and I-5 on the west and extending to El Camino Real on the east. For the most

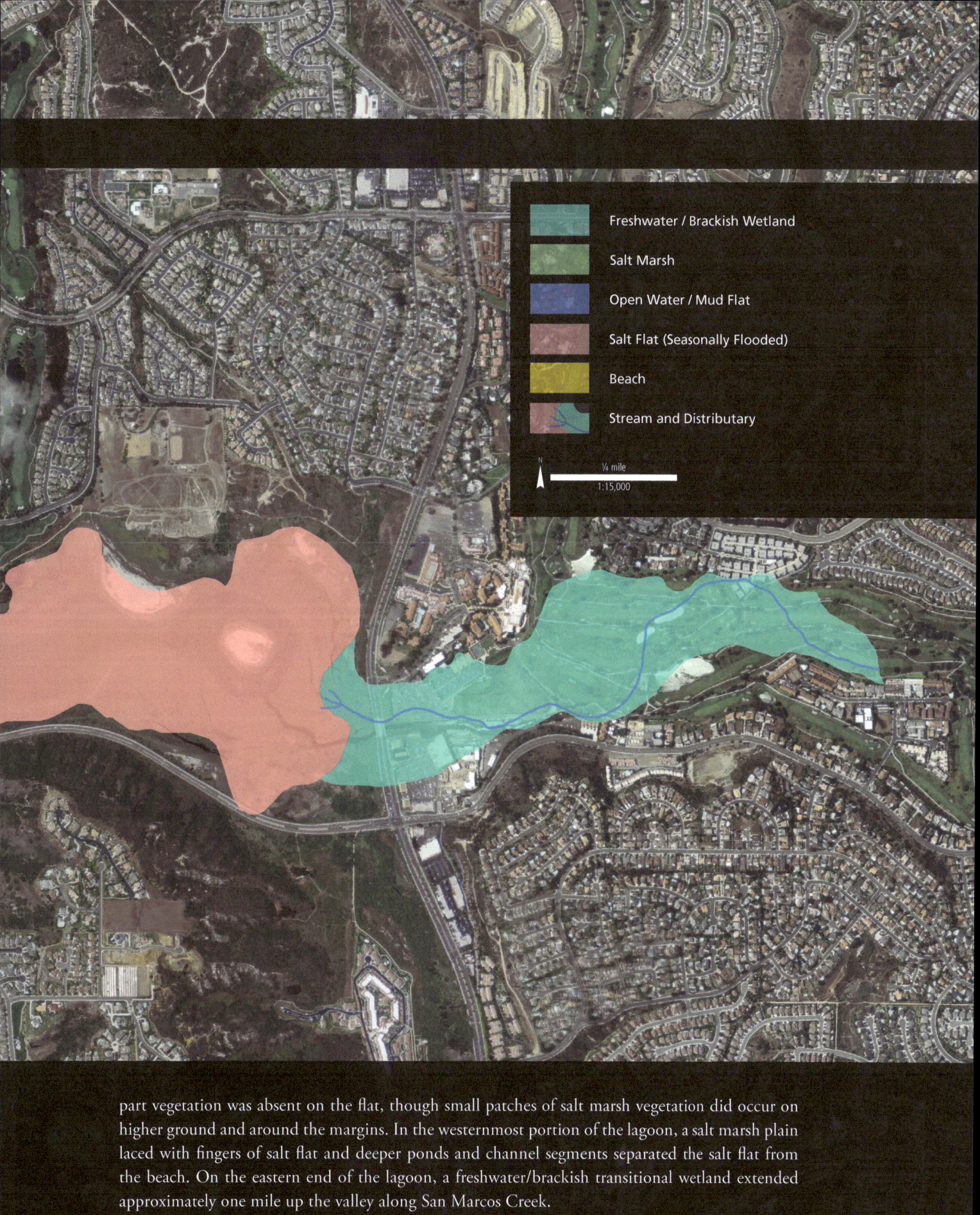

part vegetation was absent on the flat, though small patches of salt marsh vegetation did occur on higher ground and around the margins. In the westernmost portion of the lagoon, a salt marsh plain laced with fingers of salt flat and deeper ponds and channel segments separated the salt flat from the beach. On the eastern end of the lagoon, a freshwater/brackish transitional wetland extended approximately one mile up the valley along San Marcos Creek.

BEACH/INLET: A beach berm several hundred feet wide separated the lagoon from the ocean. In February of 1875, GLO surveyor Wheeler (1874-5) observed, "This marsh is shut out from the Ocean by a wall of sand and cobble stones along the beach." Seasonal breaches of the berm periodically established a tidal connection (see page 176).

SALT FLAT: Over 80% of the lagoon was covered by an unvegetated salt flat. Crespí provides the earliest documentation of this feature, writing in 1769 of "a great deal of glitter where salt must be collecting" (Crespi and Brown 2001). Seasonal cycles of flooding and evaporation contributed to the formation and maintenance of the flat: GLO surveyor Wheeler (1874-5) described how in February of 1875 the estuary was "over-flowed by the rains, but this water evaporates during the warm season, and the tract becomes almost dry." Seasonal flooding and drying of the salt flat continued to occur throughout the 20th century (Crabtree et al. 1963, Mudie et al. 1976); local residents recalled gathering salt and driving across the flat during dry periods, and boating and swimming in the lagoon when it was flooded (Lamb 1977, Haskett 1999).

SALT FLAT FINGERS INTO SALT MARSH: Narrow fingers of seasonally flooded salt flat extended into the marsh on the western side of the lagoon (see photo on page 90). Some of these fingers were connected to slightly deeper open water channels that likely held water year-round and were in turn periodically connected to the ocean (Mudie et al. 1976). This salt flat network was a persistent feature that is documented in sources spanning at least 65 years (e.g., Rodgers and McGrath 1887-8b, Unknown ca. 1925b, Unknown 1955).

Legend:

- Freshwater / Brackish Wetland
- Salt Marsh
- Open Water / Mud Flat
- Salt Flat (Seasonally Flooded)
- Beach
- Stream and Distributary

N ¼ mile
1:15,000

FRESHWATER/BRACKISH TRANSITIONAL WETLANDS: Transitional wetlands occupied the valley east of the lagoon (USGS 1891). Vegetation in more saline areas near the lagoon was likely dominated by pickleweed and saltgrass (Storie and Carpenter 1929a,b). A later report describes the marshes near tributary streams as "characterized by meadow-like cienegas, rush and 'tule' vegetation (Cat-tails *Typha latifolia*)" (Crabtree et al. 1963).

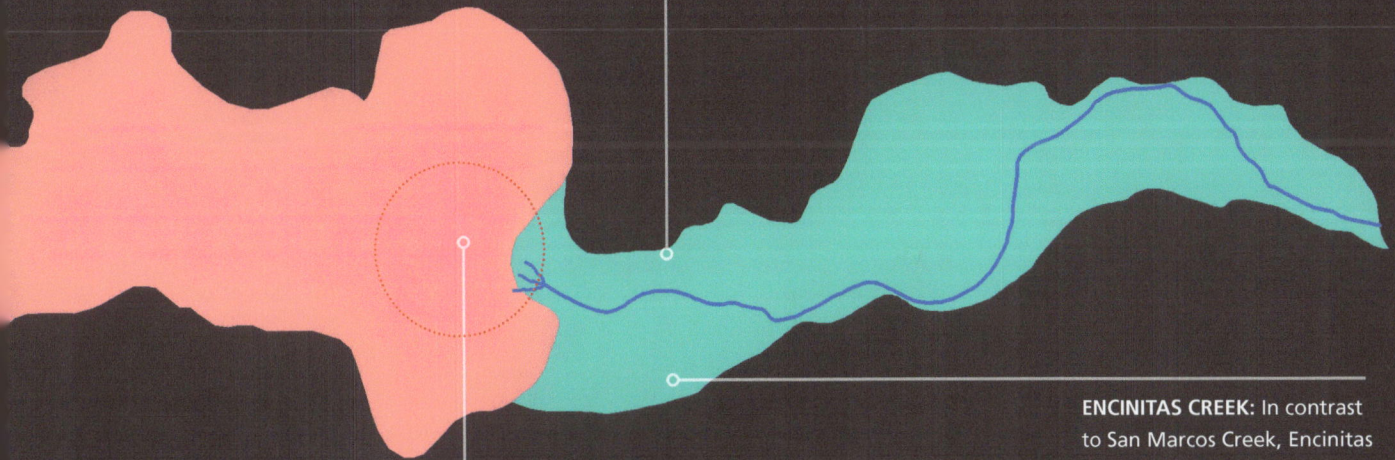

ENCINITAS CREEK: In contrast to San Marcos Creek, Encinitas Creek (not mapped) to the south flowed intermittently into the transitional wetlands upslope of the lagoon (USGS 1891).

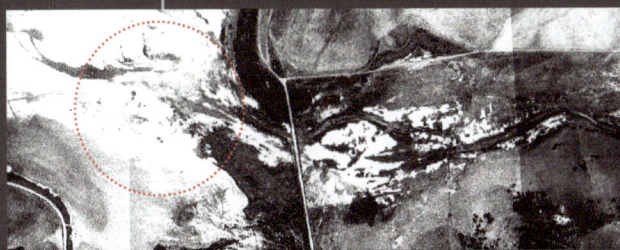

SAN MARCOS CREEK: Early sources suggest that San Marcos Creek maintained at least some surface water during the dry season through the canyon downstream of San Marcos before spreading into a freshwater/brackish wetland at the upstream end of the lagoon (USGS 1891, 1901b). However, records from Spanish explorers' journals indicate that flow was relatively limited through the canyon during the summer: though they found "springs of water" in the canyon in July 1769, it was necessary to "[dig them] out a little [so] the animals could drink" (Crespí and Bolton 1927). The term *batequi* itself suggests the presence of subsurface water (see note on page 80).

Beach/Inlet: USDC ca. 1840b, courtesy of The Bancroft Library; Freshwater/Brackish Transitional Wetlands: USGS 1891; San Marcos Creek: San Diego County 1928.

> I once watched a bean thresher, pulled by four horses, cross the dry bed of the lagoon... In wet years, we kids would row across in a boat, and dive off a small wooden dock that was there then.
>
> —RICHARD LYMAN IN HASKETT 1999, REFERRING TO BATIQUITOS LAGOON IN THE EARLY 1900S

The boundary between the salt marsh and salt flat is visible near the center of this ca. 1925 photograph, with a large expanse of salt flat extending to the east. Narrow fingers of salt flat are visible extending into the marsh. Though the central portion of the lagoon is still flooded at the time of the photograph, a lighter-colored "bathtub ring" where the salt flat is drying out can be seen around the margin. (photo #90:18138-667, courtesy of San Diego History Center)

(top right) Leucadia is shown in the foreground of this 1955 photograph, with Batiquitos Lagoon visible in the background. The pre-railroad channel configuration can still be distinguished, but water has backed up against both sides of the railroad berm. (Unknown 1955, courtesy of San Dieguito Heritage Museum)

(bottom right) The central portion of Batiquitos Lagoon is visible in the upper left side of this 1932 photograph. The salt flat appears to be partially dried out at the time this photograph was taken. (photo #79:741-1531, courtesy of San Diego History Center)

BATIQUITOS LAGOON: Change Over Time

EARLY CHANGES TO BATIQUITOS LAGOON INCLUDED CONSTRUCTION OF THE RAILROAD BERM AND THE COAST HIGHWAY, WHICH ALTERED FLOW PATTERNS IN THE OPEN WATER CHANNELS. San Marcos Dam was constructed in 1952, resulting in decreased freshwater and sediment inputs to the lagoon (Gayman 1978b). In the 1960s and 70s, sewage effluent was released into the lagoon, altering salinity levels and contributing to water quality problems (County of San Diego 1970, Mudie et al. 1976).

Despite these manipulations, the habitat distribution in Batiquitos Lagoon remained somewhat persistent throughout much of the 20th century. As late as 1980, salt flat and mud flat still occupied nearly 60% of the estuary (Mudie et al. 1976, Meyer 1980). Reports from the 1960s and 70s described the lagoon as "a large expanse of salt-incrusted playa" (Miller 1966) or a "barren salt flat which is inundated by 4 to 6 inches of water during the rainy season" (Mudie et al. 1976).

The Batiquitos Lagoon Enhancement Project was completed in 1996, transforming the lagoon into a continuously tidal, open water estuary. The project involved dredging of the lagoon basin and construction of jetties to stabilize the mouth of the lagoon (Merkel & Associates, Inc. 2009). Today, the estuarine portion of Batiquitos Lagoon is made up of 49% open water/mud flat, 48% salt marsh, and 3% salt flat.

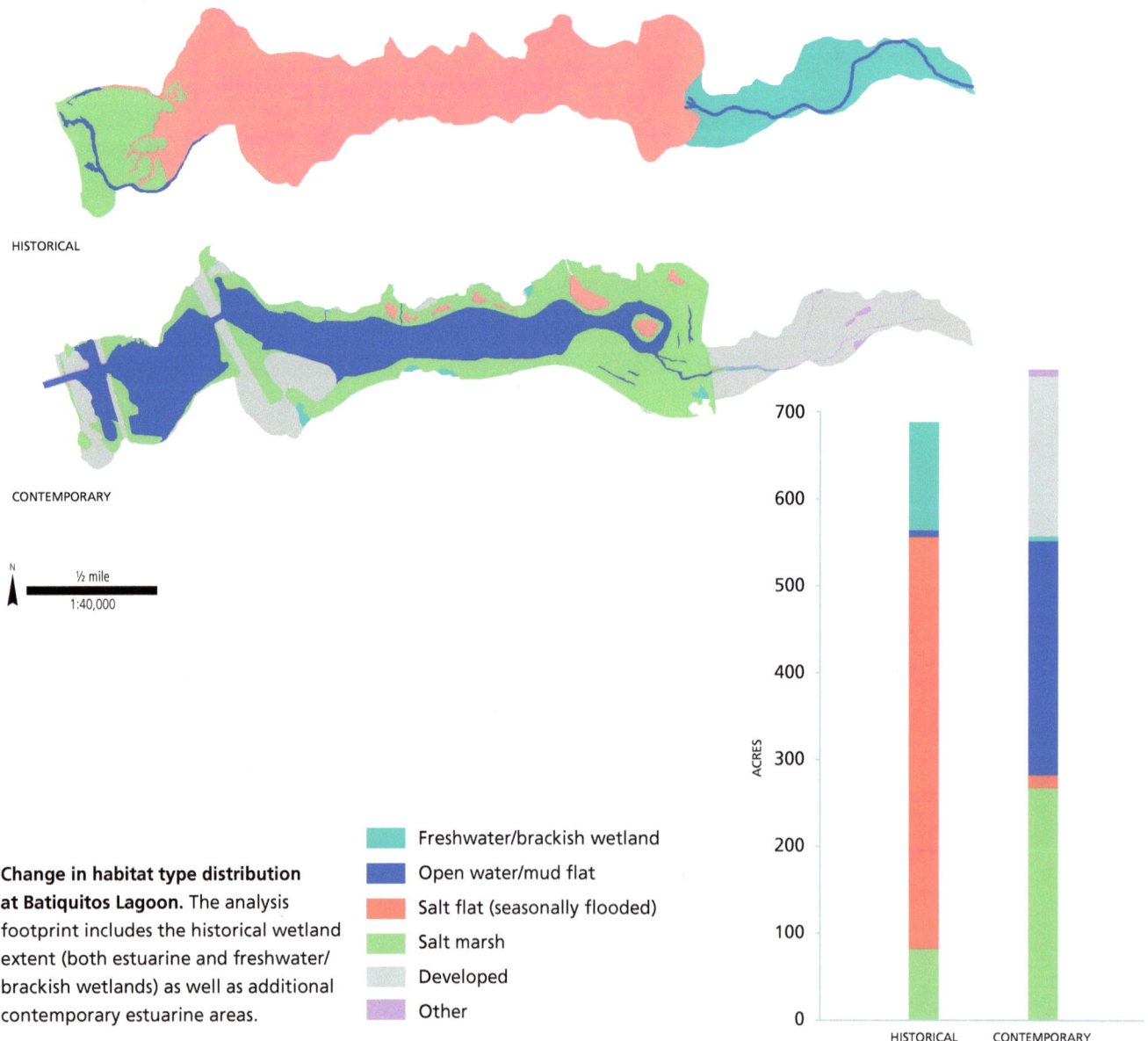

HISTORICAL

CONTEMPORARY

½ mile
1:40,000

Change in habitat type distribution at Batiquitos Lagoon. The analysis footprint includes the historical wetland extent (both estuarine and freshwater/brackish wetlands) as well as additional contemporary estuarine areas.

Legend:
- Freshwater/brackish wetland
- Open water/mud flat
- Salt flat (seasonally flooded)
- Salt marsh
- Developed
- Other

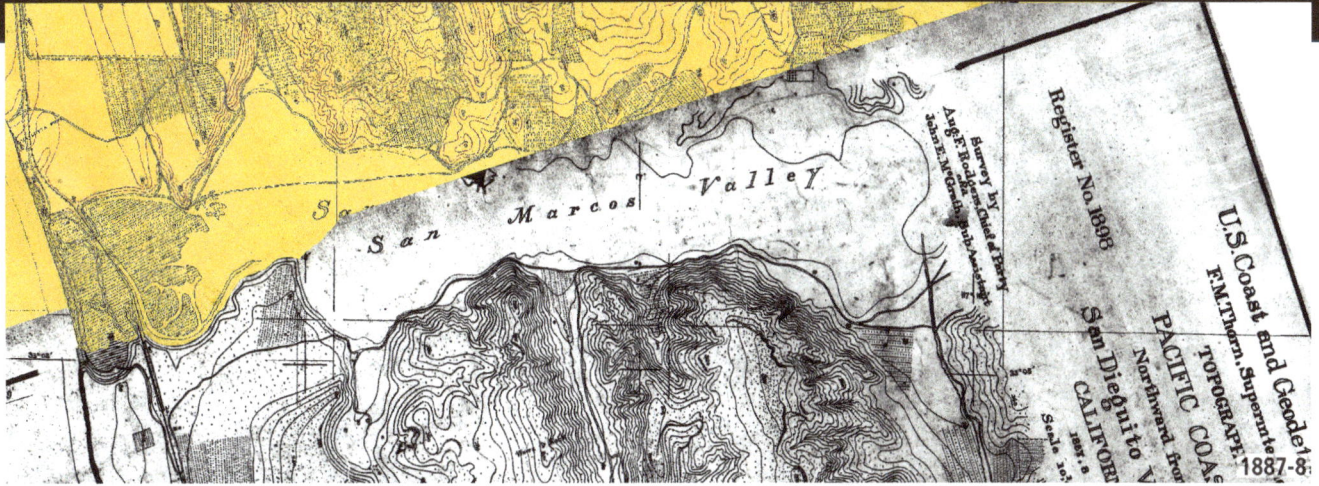

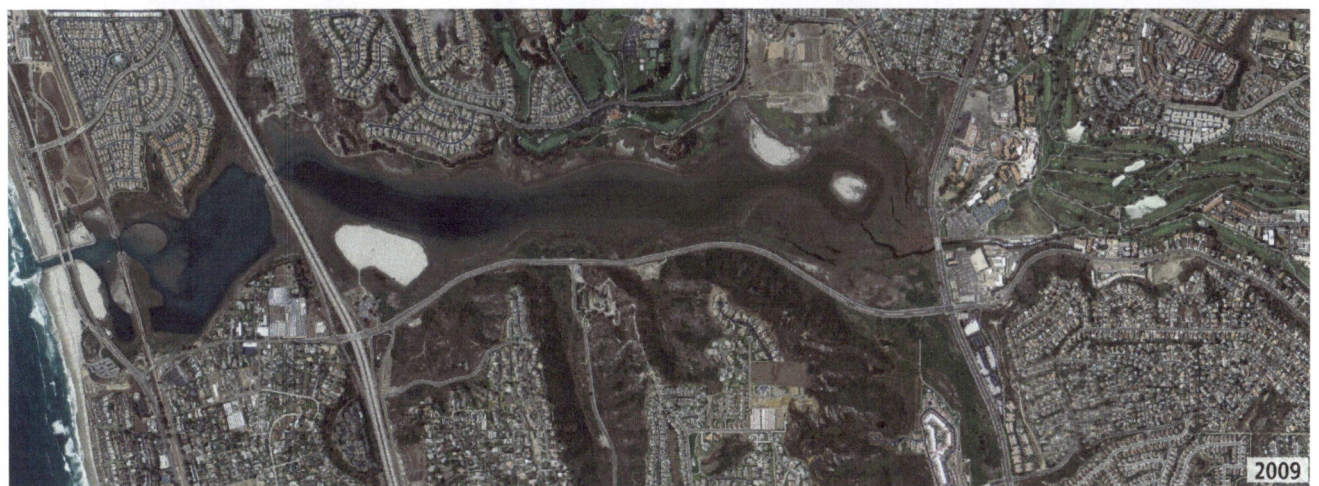

Construction of the railroad berm in 1881-2 bisected the salt marsh at the western end of Batiquitos Lagoon, as can be seen on the T-sheet **(top)**. Salt flat still occupied the vast majority of the lagoon at the time of the 1928 aerial photos **(middle)**, and continued to be a dominant habitat type until as late as 1980. Today, open water/mud flat and salt marsh are the dominant habitat types at Batiquitos Lagoon **(bottom)**. (Rodgers and McGrath 1887-8a,b, San Diego County 1928, NAIP 2009)

7. SAN ELIJO LAGOON

San Alejo was the name given to the canyon upstream of Batiquitos Lagoon, where the Portolá Expedition camped on July 16, 1769, the night before the feast day of St. Alexius. The name later became associated with the current San Elijo Lagoon (Gudde and Bright 1998).

SAN ELIJO LAGOON IS A 600 ACRE ESTUARY LOCATED BETWEEN THE TOWNS OF ENCINITAS AND SOLANA BEACH. It supports a diverse array of habitats including coastal salt marsh, tidal mud flats, freshwater marsh, and salt pannes (AECOM 2012). Transportation corridors divide the lagoon into three basins: Highway 101 runs along the coast, the Santa Fe Railroad lies a short ways to the east, and Interstate 5 crosses through the middle of the lagoon. The San Elijo Lagoon Ecological Reserve was designated in 1983, and is managed jointly by the State of California, County of San Diego Parks & Recreation, and the San Elijo Lagoon Conservancy. The lagoon is fed by Escondido and La Orilla creeks, and its watershed covers 85 square miles.

San Elijo Lagoon. (photo by Sean Baumgarten, January 2013)

Timeline: San Elijo Lagoon

1875	One of the first farmers in the area, Hector MacKinnon, settles just north of San Elijo Lagoon.
1880s-1940s	Dikes and levees constructed in San Elijo Lagoon to make duck ponds, sewage settling ponds, and roads.
1881-82	California Southern Railroad line constructed between National City and Oceanside.
1884	Colony Olivenhain organized, converting much of Rancho Las Encinitas (partially within San Elijo Lagoon's watershed) into olive orchards.
1895	Lake Wohlford Dam constructed.
Early 1900s	Water diverted from San Luis Rey River to Escondido Creek for irrigation.
1912-15	Pacific Coast Highway constructed.
1920s	The San Dieguito and Santa Fe Irrigation Districts begin supplying water from Lake Hodges (in the San Dieguito River watershed) to farms and homes in the San Elijo Lagoon watershed.
1940-73	Sewage effluent discharged into San Elijo Lagoon and Escondido Creek.
1965-67	Interstate 5 constructed.
1971	Lake Dixon Dam constructed.
1981	Water management infrastructure, including dikes, gates, and spillways, constructed in the lagoon's east basin.
1987	San Elijo Lagoon Conservancy established.

Sources: Adams et al. 1912, Holmes and Pendleton 1918, California Department of Public Works 1949, Goodwin et al. 1992, Welker and Patton 1995, County of San Diego 1996, Tucker and Bujkovsky 2009

Solana Beach, 1923. Agriculture around the community of Solana Beach dominated the south shore of San Elijo Lagoon in the early 20th century. Avocado orchards occupy the foreground in this photograph; the lagoon is just visible in the upper right corner. (photo courtesy of San Dieguito Heritage Museum)

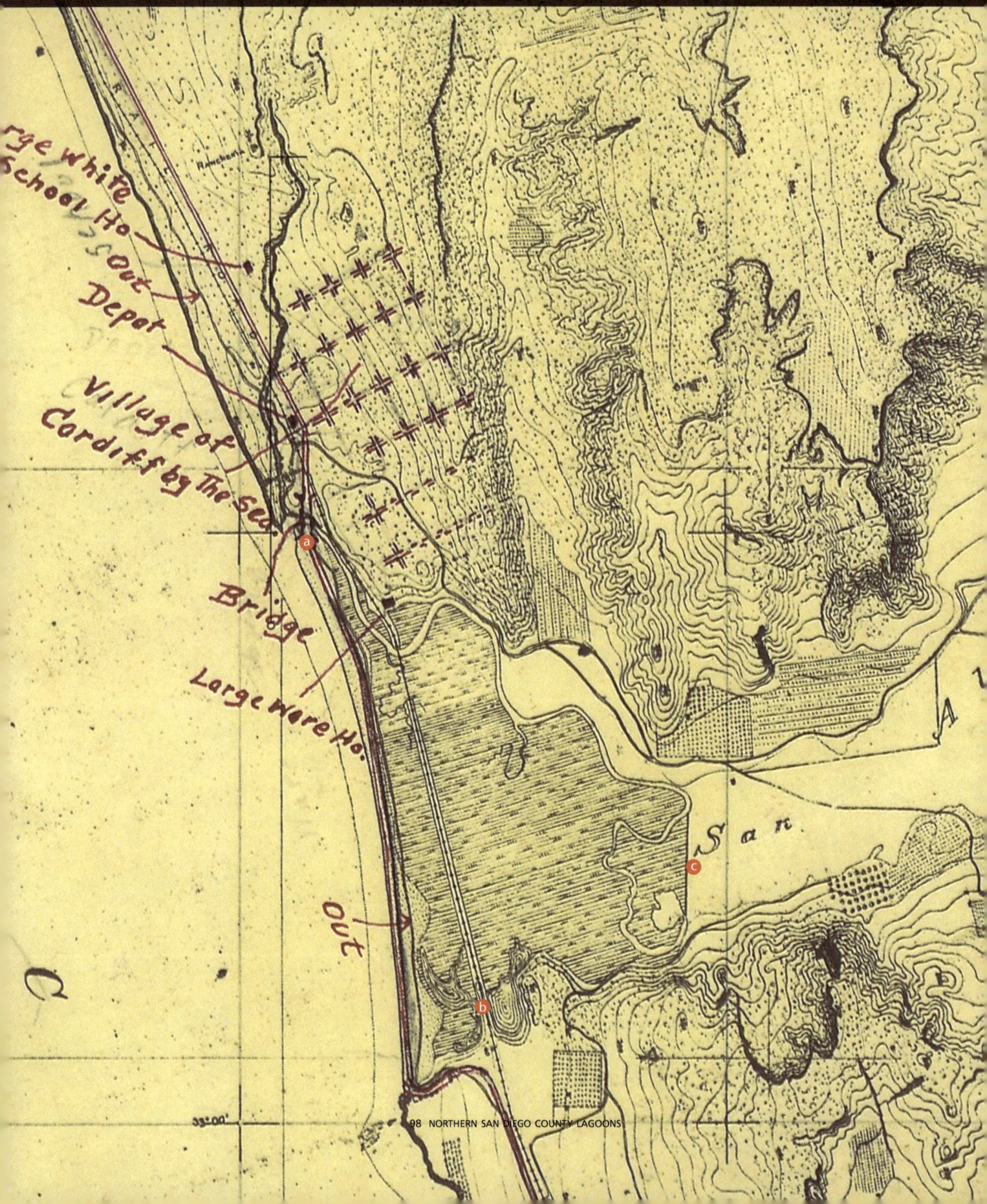

rge White

School Ho — Out

Depot

Village of

Cardiff by the Sea

Bridge

Large Ware Ho

Out

Out

Out

San

San Elijo Lagoon as shown on the T-sheet, surveyed August to December 1887 and May to November 1888. (Rodgers and McGrath 1887-8a; courtesy of NOAA)

a Unlike the northern three lagoons, the inlet at San Elijo Lagoon (in the northwest corner) is depicted as open at the time of the T-sheet survey.

b Open water ponds mark the location of an alternate inlet on the south side of the lagoon.

c A man-made dike in the middle of the lagoon separated the salt marsh on the western side of the lagoon from the salt flat to the east.

d As at Buena Vista Lagoon, numerous roads criss-cross the salt flat, indicating paths for dry-season travel.

e The inland boundary of the southeastern arm of the lagoon is not clearly defined on the T-sheet.

¼ mile
1:15,000

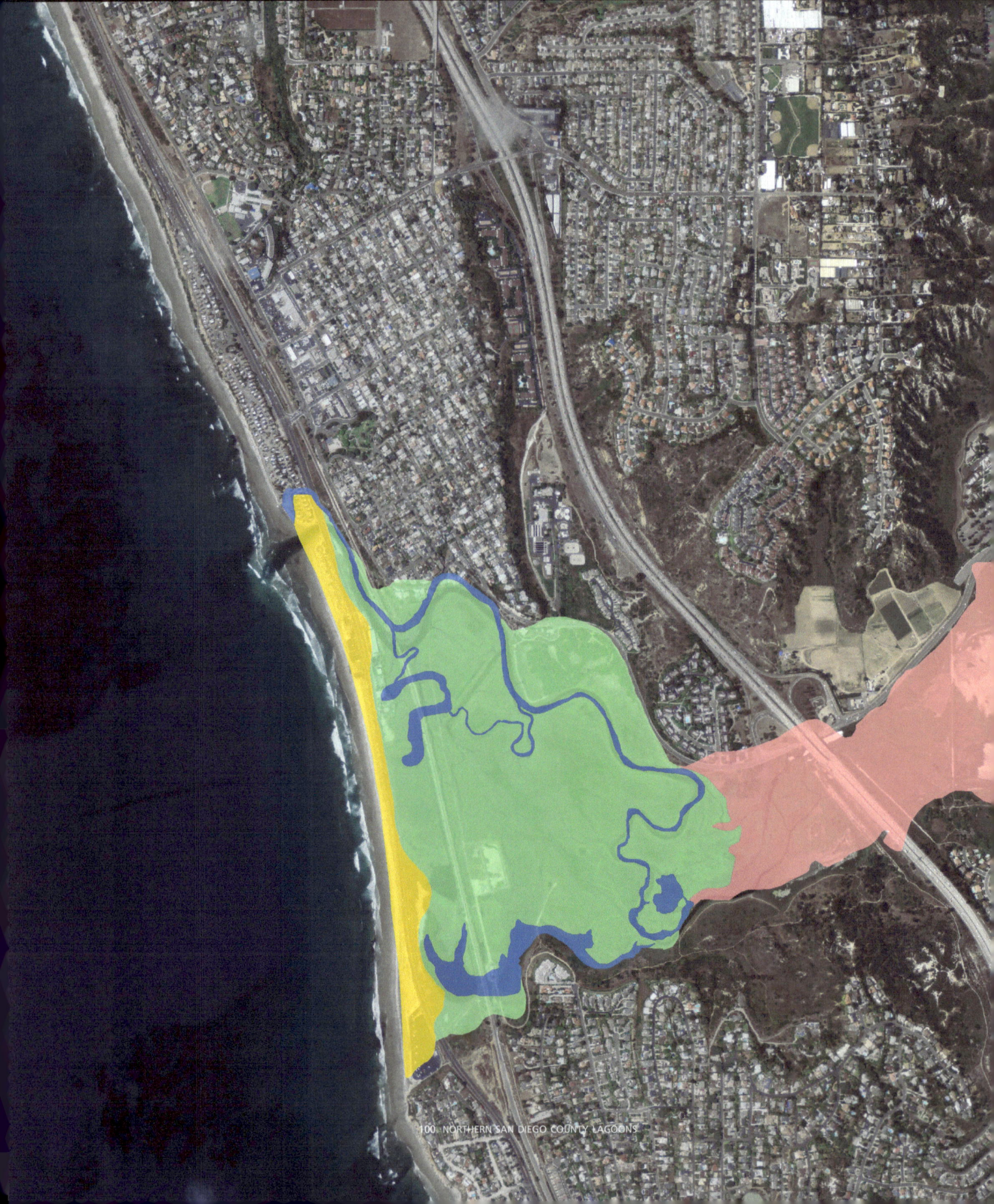

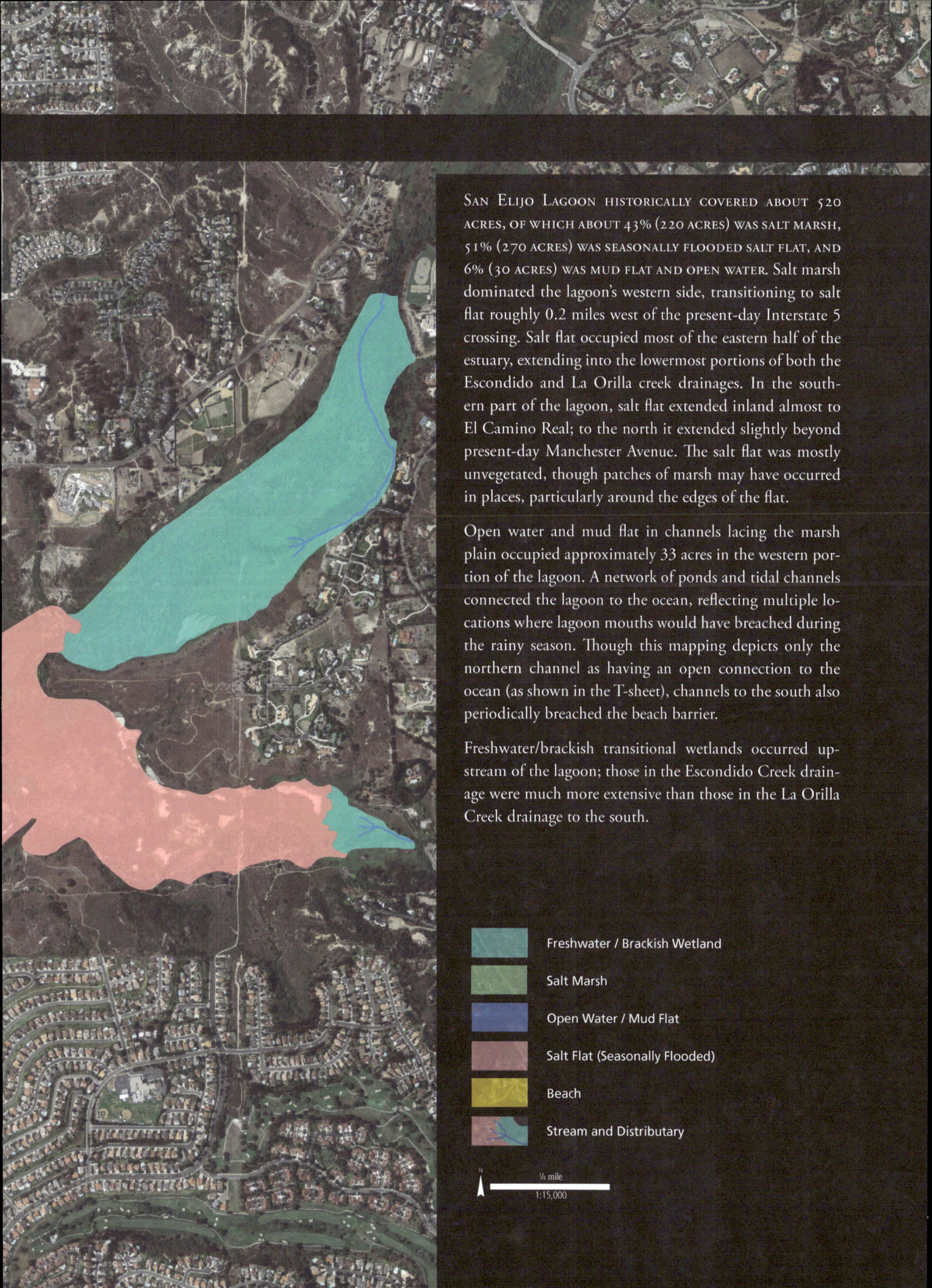

SAN ELIJO LAGOON HISTORICALLY COVERED ABOUT 520 ACRES, OF WHICH ABOUT 43% (220 ACRES) WAS SALT MARSH, 51% (270 ACRES) WAS SEASONALLY FLOODED SALT FLAT, AND 6% (30 ACRES) WAS MUD FLAT AND OPEN WATER. Salt marsh dominated the lagoon's western side, transitioning to salt flat roughly 0.2 miles west of the present-day Interstate 5 crossing. Salt flat occupied most of the eastern half of the estuary, extending into the lowermost portions of both the Escondido and La Orilla creek drainages. In the southern part of the lagoon, salt flat extended inland almost to El Camino Real; to the north it extended slightly beyond present-day Manchester Avenue. The salt flat was mostly unvegetated, though patches of marsh may have occurred in places, particularly around the edges of the flat.

Open water and mud flat in channels lacing the marsh plain occupied approximately 33 acres in the western portion of the lagoon. A network of ponds and tidal channels connected the lagoon to the ocean, reflecting multiple locations where lagoon mouths would have breached during the rainy season. Though this mapping depicts only the northern channel as having an open connection to the ocean (as shown in the T-sheet), channels to the south also periodically breached the beach barrier.

Freshwater/brackish transitional wetlands occurred upstream of the lagoon; those in the Escondido Creek drainage were much more extensive than those in the La Orilla Creek drainage to the south.

Freshwater / Brackish Wetland

Salt Marsh

Open Water / Mud Flat

Salt Flat (Seasonally Flooded)

Beach

Stream and Distributary

¼ mile
1:15,000

INTERMITTENTLY TIDAL PONDS AND CHANNELS: Maps from the early 1880s – prior to the construction of the California Southern Railroad – reveal the open water ponds shown in the T-sheet to be inlet locations at different points in time. The construction of the railroad in 1881-2 bisected the channels with a berm, cutting off these alternate pathways for fluvial and tidal exchange. Despite this, these features persisted as deep ponds well into the 20th century; the southernmost pond is still present today (see pages 106-107).

SALT FLAT: Salt flat occupied the eastern portion of San Elijo Lagoon, covering just over half of the estuary's total area. Surveyor Rodgers (1887-8a) described the salt flat as a mile-long "shallow lagoon" in the winter, while in the dry season it transformed into a "glistening white alkaline plain, marked only by the dark lines of travel, which turn up the subjacent black loam or A-do-be." The salt flat was used by native residents, who gathered salt from the flat (Harrington 1925; see pages 104-105).

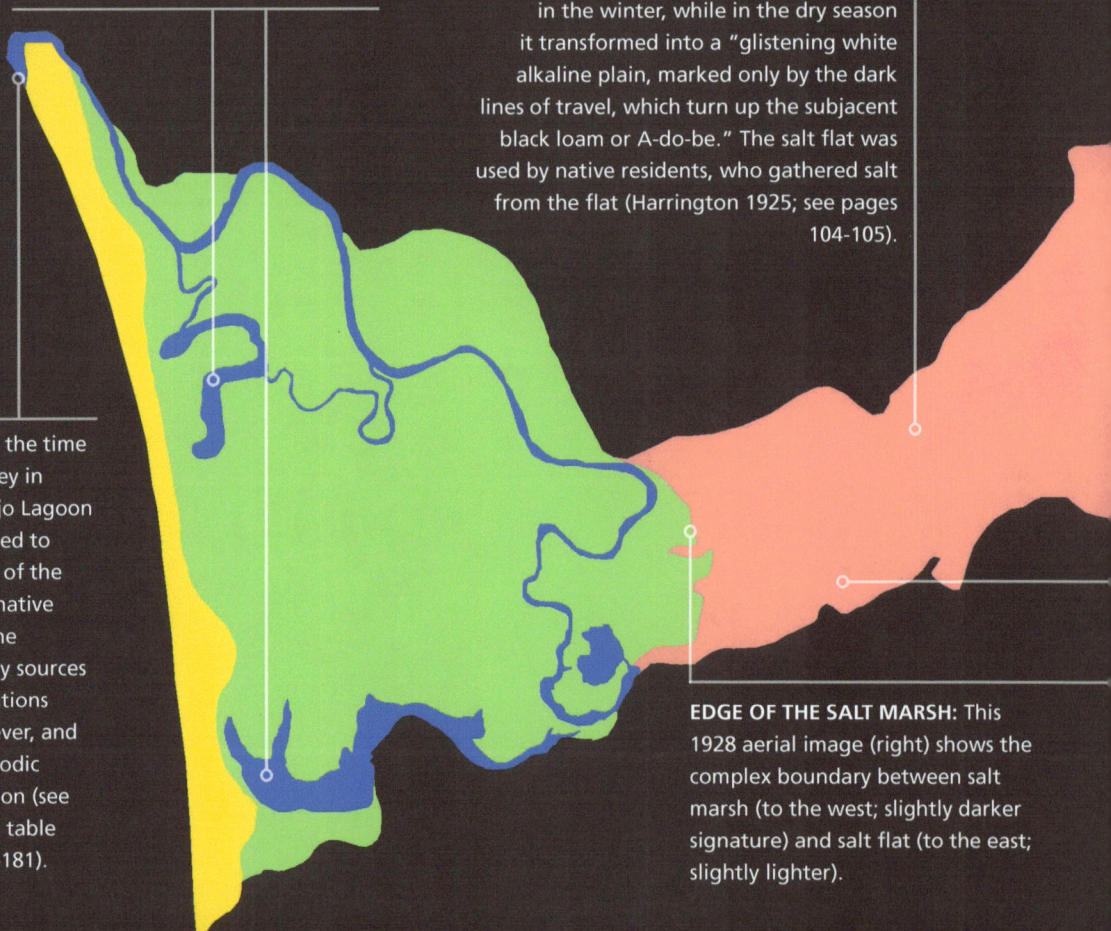

LAGOON INLET: By the time of the T-sheet survey in 1887-8, the San Elijo Lagoon inlet was constrained to the northern edge of the estuary, with alternative routes cut off by the railroad berm. Early sources show multiple locations for the inlet, however, and document the periodic closure of the lagoon (see pages 106-107 and table 11.2 on pages 180-181).

EDGE OF THE SALT MARSH: This 1928 aerial image (right) shows the complex boundary between salt marsh (to the west; slightly darker signature) and salt flat (to the east; slightly lighter).

FRESHWATER/BRACKISH WETLANDS: Freshwater/brackish wetlands were found upslope of San Elijo Lagoon, including this extensive area surrounding Escondido Creek. Mapping from the turn of the last century (USGS 1901a) suggests these wetlands may have extended even further inland along the creek than mapped here.

■ (teal)	Freshwater / Brackish Wetland
■ (green)	Salt Marsh
■ (blue)	Open Water / Mud Flat
■ (salmon)	Salt Flat (Seasonally Flooded)
■ (yellow)	Beach
■	Stream and Distributary

N

¼ mile
1:15,000

It is apparent that during the dry seasons the mouth of the lagoon becomes entirely closed by tide action, and as the lagoon fills up with fresh water during the rainy seasons the mouth is again opened and it becomes tidal until this action is repeated.

—KNOX 1934-5B

HEAD OF ESTUARY: GLO surveyor C.F. Hoffman, surveying San Elijo Lagoon in January 1869, noted that here he was near the "head of inlet coming from sea" (Hoffman 1869) – that is, the easternmost edge of the estuary.

INUNDATED SALT FLAT: On October 14, 1880, GLO surveyor A.P. Hanson noted a survey post "falling in water about 1 ft. deep" on the shallowly flooded salt flat (Hanson 1880). The origin of this water is unclear; it may have represented tidal flooding (if the inlet were open at the time) or impounded water (if the mouth were closed).

Ponds and Channels: Osgood 1881b, courtesy of California State Railroad Museum; Salt Flat: photo #91-34493, courtesy of National Anthropological Archives, Smithsonian Institution; Edge of Salt Marsh: San Diego County 1928.

SAN ELIJO LAGOON: Salt Flat

(right) Dry salt flat at San Elijo Lagoon in August 1925, looking west. Fringing patches of salt marsh can be seen around the edges of the flat. In the caption for the original image, the photographer (ethnographer John P. Harrington) noted that the native inhabitants "used to get salt off the two points of hill jutting out at left of gap to sea." (photo #91-34493, courtesy of National Anthropological Archives, Smithsonian Institution)

(below) Ethnographer and linguist John Peabody Harrington (1884-1961). (courtesy of National Anthropological Archives, Smithsonian Institution)

SAN ELIJO LAGOON: Spatial Variability in Inlet Location

THE CONSTRUCTION OF THE CALIFORNIA SOUTHERN RAILROAD BERM ACROSS SAN ELIJO LAGOON IN 1881-2 BLOCKED TID-AL-FLUVIAL EXCHANGE EVERYWHERE EXCEPT THROUGH A SINGLE CHANNEL RUNNING UNDER THE RAILROAD BRIDGE ON THE NORTHERN SIDE OF THE BERM. The inlet was confined to the northwestern corner of the lagoon, as shown on the T-sheet (Rodgers and McGrath 1887-8a) and in the historical aerials (San Diego County 1928); the lagoon inlet occupies the same general location today (see images below).

While this northern inlet marked a historically persistent inlet location, additional sources – in particular, a series of railroad maps from the early 1880s – provide evidence for the existence of multiple additional inlet locations before the railroad berm constricted the inlet (see facing page). Though these maps were created as part of the route assessment for the California

The 1887-8 T-sheet, 1928 aerial photographs, and 2009 aerial photographs of San Elijo Lagoon all show an open inlet at the northern edge of the estuary. The inlet was confined to this general location by the construction of the California Southern Railroad berm in 1881-2 **(left)**. Remnants of several alternative inlet locations that would have been activated historically (see facing page) are still visible in the modern aerial photos in the form of ponds and channels within the marsh plain. In the 1928 photograph **(center)**, remnants of both a central and southern inlet are visible (circled), while by 2009 **(right)** only the southern inlet is still marked by a pond (circled); traces of the central inlet have been removed. (Rodgers and McGrath 1887-8a, San Diego County 1928, NAIP 2009)

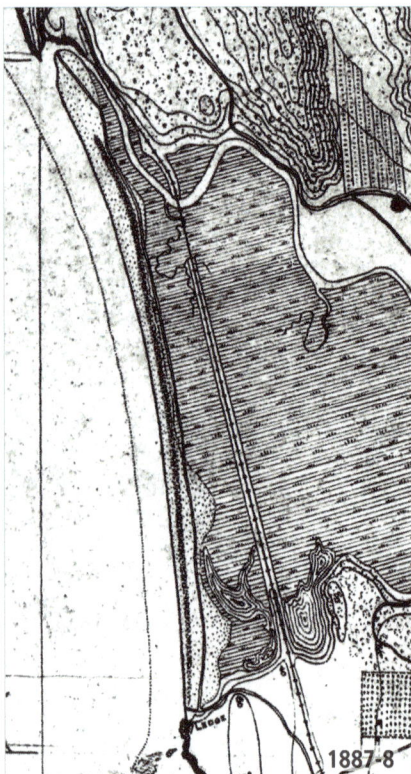

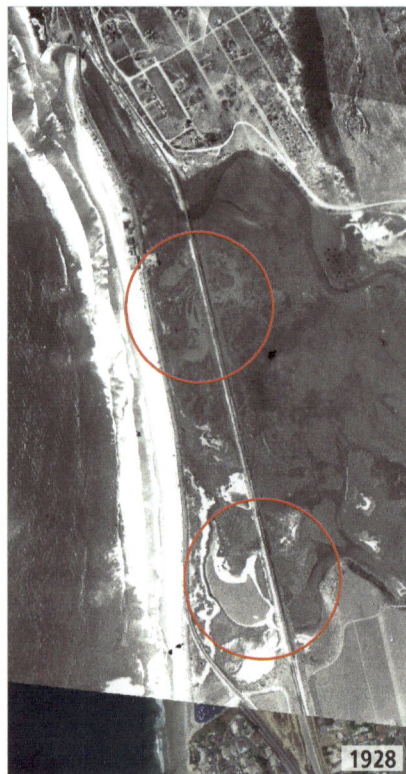

Southern Railroad, they were surveyed prior to the railroad's construction. It is apparent from this suite of maps that the mouth of San Elijo Lagoon breached in different locations along the beach barrier at different times. Following large flood events, multiple inlets may have been activated simultaneously.

Although the additional inlet locations documented in the early 1880s were no longer activated after construction of the railroad berm bisected channels leading to the inlets, relicts of these features persisted as ponds and channels throughout the 20th century (see images below and photograph on pages 108-109). These "fossilized" inlet and channel remnants were some of the deepest portions of the estuary, retaining surface water in the dry season longer than other parts of the lagoon (USGS [1891]1898).

Three early 1880s railroad maps, overlaid with the historical synthesis mapping for context, show multiple inlet locations at San Elijo Lagoon. The map on the **bottom left** (ca. 1881, season unknown) offers no indication of a northern inlet, and instead shows a single open inlet at the southern edge of the lagoon corresponding to the location of the open water pond shown in the historical mapping. The map at **bottom center** (February 1881) also shows a southern inlet (though here shown at least partially closed), and depicts an additional middle inlet extending from the U-shaped channel shown in the synthesis mapping; the northern route (now the main inlet) is not connected to the ocean. Finally, the map at **bottom right** (ca. 1881, season unknown) shows a southern and northern inlet, both open (the northern inlet is slightly displaced compared to the location shown on the T-sheet and historical aerials). The precise meaning of this suite of maps is ambiguous: given the proximity in date and similarity in purpose of the maps, it is not clear if these differences represent actual changes in inlet location over time, or different surveyors or cartographers depicting similar inlet conditions in different ways. Furthermore, it is not clear whether the railroad bridge over the northern channel reinforced what was already a primary inlet, or whether an alternate channel route was more frequently occupied prior to the construction of the bridge and berm. Regardless, these maps clearly illustrate spatial variability in inlet location and highlight the dynamic nature of the inlet prior to the railroad. (Sources: Unknown ca. 1881a, Osgood 1881a, Unknown ca. 1881b)

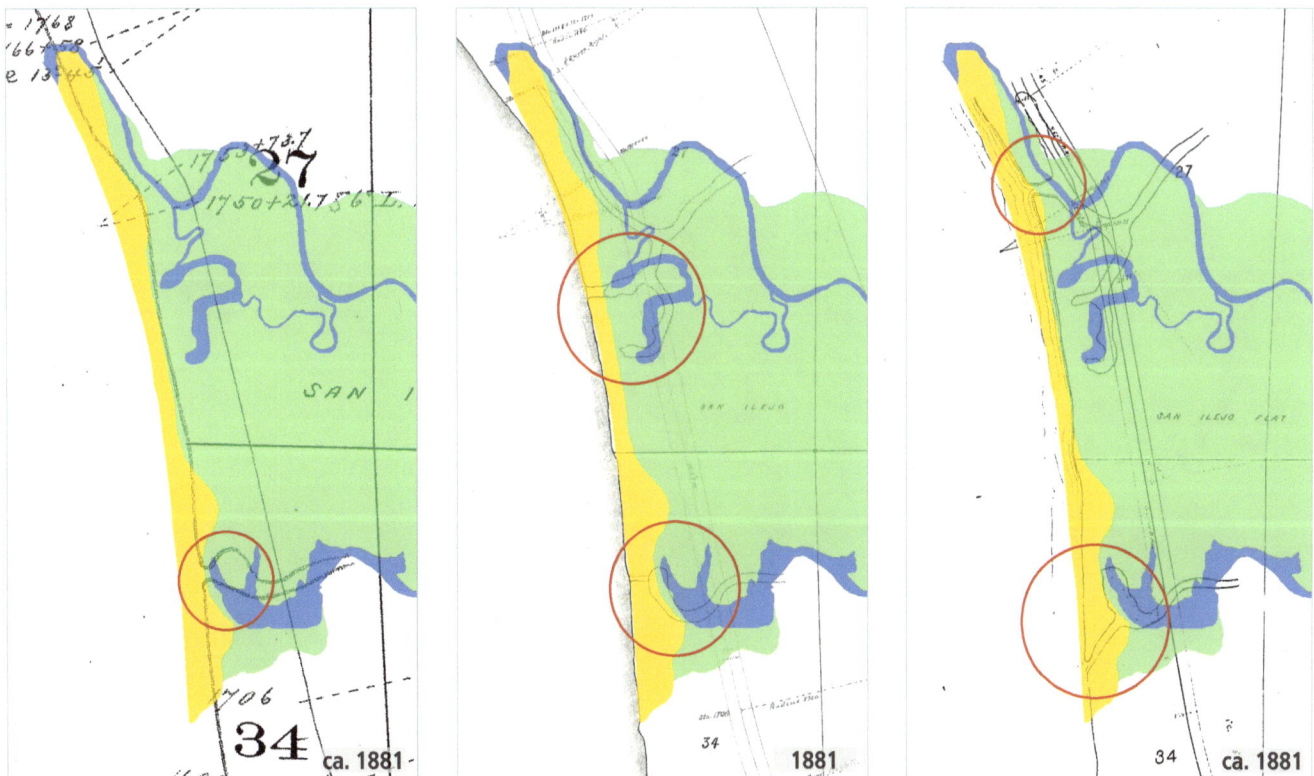

The impact of the railroad berm (visible here in the foreground, behind Highway 101) on water circulation and habitat distributions within the lagoon can be observed in this aerial photo from September 1954. The berm bisected the salt marsh, severed the connection between the channel remnants visible in the foreground and the remainder of the channel network, and confined the inlet to the northern side of the lagoon. Both the channel remnants in the foreground and the salt flat in the background appear dry at the time the photograph was taken. (Collection 87-26, USA-C1 54 15/3, courtesy of Scripps Institution of Oceanography Archives, UC San Diego)

RELATIVE TO THE OTHER LAGOONS IN THE REGION, SAN ELIJO LAGOON HAS RETAINED SUBSTANTIAL COMPONENTS OF ITS HISTORICAL HABITAT PROFILE. Many components of the channel network are still intact, and salt marsh still exists in over half of the areas where it existed historically.

However, major transformations have also taken place. Of the approximately 270 acres of salt flat that formerly occupied the eastern portion of the lagoon, less than 50 acres (~18%) remain today. Much of the salt flat has been replaced by freshwater/brackish wetlands or salt marsh, driven by a combination of land use changes including construction of highway and railroad bridges, sewage effluent discharge, and increases in urban and agricultural runoff (Welker and Patton 1995). Freshwater/brackish wetlands have increased in extent by nearly 60%, while salt marsh has increased by approximately 27%.

Large portions of the salt flat persisted well into the 20th century. The flat is visible in the 1928 historical aerial imagery, covering a nearly identical extent as that shown on the 1887-8 T-sheet (see facing page). In the middle of the 20th century, the boundary between the western salt marsh and eastern salt flat was also still quite distinct, especially during the dry season. Accounts of the lagoon from this time period capture the seasonally dry nature of the salt flat: Purer (1942) describes the lagoon as having "little open water, particularly in summer." Extreme fluctuations in salinity occurred from winter to summer (Purer 1942, Carpelan 1969). As late as 1995, about 30% of the eastern basin supported seasonally flooded salt flats (Welker and Patton 1995).

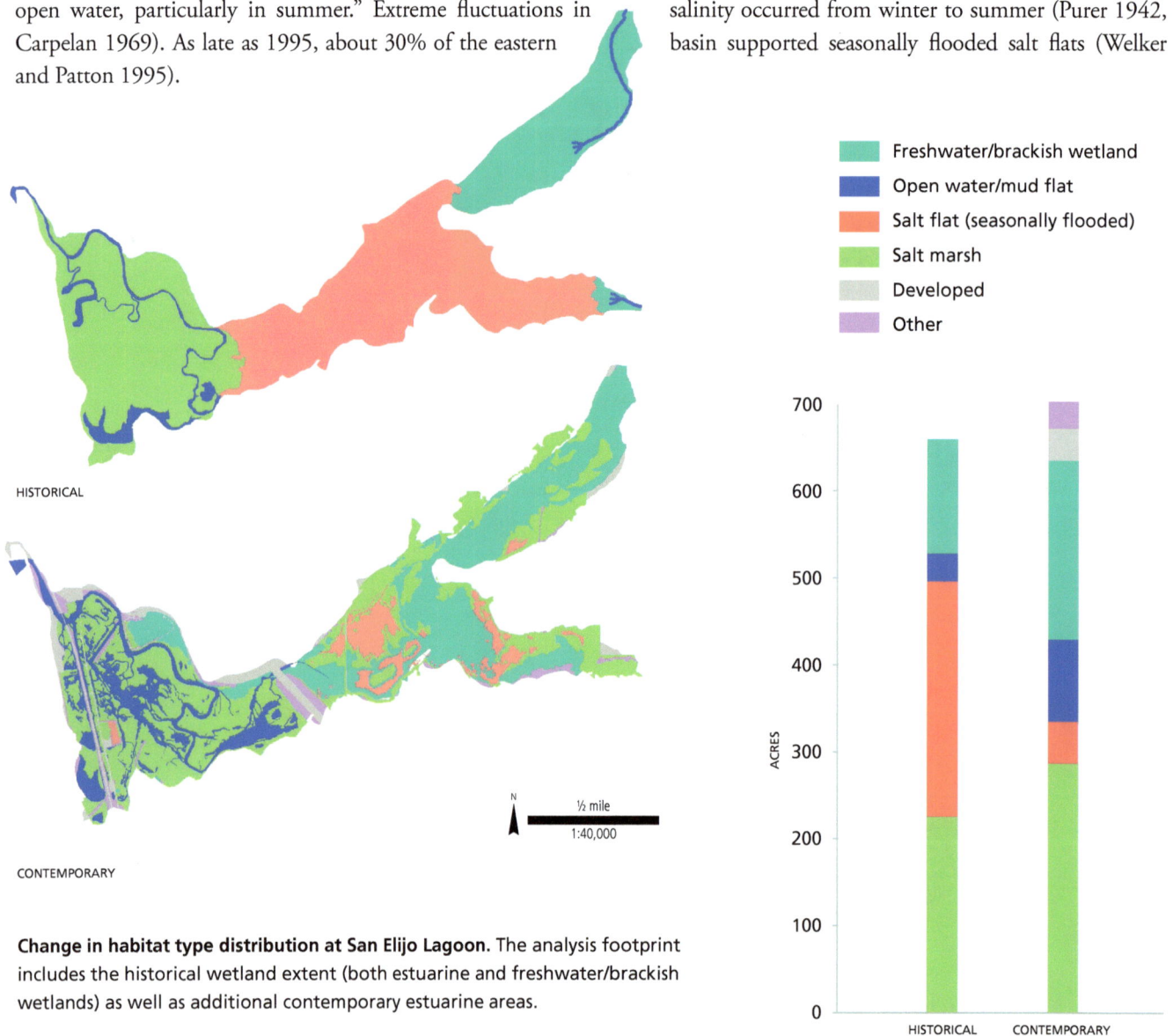

HISTORICAL

CONTEMPORARY

½ mile
1:40,000

Change in habitat type distribution at San Elijo Lagoon. The analysis footprint includes the historical wetland extent (both estuarine and freshwater/brackish wetlands) as well as additional contemporary estuarine areas.

Freshwater/brackish wetland
Open water/mud flat
Salt flat (seasonally flooded)
Salt marsh
Developed
Other

1887-8

1928

2009

N
½ mile
1:40,000

At the time the T-sheet was made, a dike separated the salt marsh on the western side of the lagoon from the salt flat to the east **(top)**. The 1928 aerial photos **(middle)** appear to show a more gradual transition between these two habitat types. The contemporary lagoon still supports extensive areas of salt marsh in the western basin, and several remnant patches of salt flat in the eastern basin are visible in the 2009 aerial photos **(bottom)**. Freshwater/brackish marsh has expanded significantly within the eastern portion of the lagoon. (Rodgers and McGrath 1887-8a, San Diego County 1928, NAIP 2009)

8. SAN DIEGUITO LAGOON

San Dieguito Lagoon (photo by Sean Baumgarten, January 2013)

SAN DIEGUITO LAGOON (FORMERLY ALSO KNOWN AS DEL MAR LAGOON) IS LOCATED IN THE CITIES OF DEL MAR AND SAN DIEGO, JUST SOUTH OF SOLANA BEACH. The lagoon is the terminus of the San Dieguito River, which drains an approximately 345 square mile watershed – much larger than the watersheds of the other lagoons studied. Beginning in the early 20th century, large portions of the marsh plain were filled for construction of roads and highways, an airfield, the Del Mar Fairgrounds, and a shopping center. These developments greatly reduced the estuary's area from its historical extent, though restoration efforts in recent years have compensated for this loss to some degree. The lagoon currently covers about 500 acres, the majority of which is salt marsh.

Interstate 5 divides the modern estuary into two main sections: the portion west of I-5 and south of Jimmy Durante Boulevard occupies approximately 230 acres, while the portion east of I-5 occupies approximately 260 acres. A small area of marsh also occurs between the railroad and Camino Del Mar (Highway 101).

Timeline: San Dieguito Lagoon

1845	Rancho San Dieguito granted to Juan María Osuna, the first alcalde (mayor) of the Pueblo of San Diego.
1853	East San Pasqual Ditch, one of the earliest irrigation projects in the San Dieguito watershed, constructed.
1881-82	California Southern Railroad line constructed between National City and Oceanside.
1887	West San Pasqual Ditch constructed.
1895	Lake Wohlford Dam constructed.
Early 1900s	Water pumped from San Dieguito river bed to supply water for Del Mar.
1910s	Widespread cultivation of sugar beets and lima beans in the river valley.
Early 1900s-70s	Portions of the lagoon filled for roads, highways, and development projects.
1912-15	Pacific Coast Highway constructed.
1918	Lake Hodges Dam constructed.
1935-37	Del Mar Fairgrounds constructed.
1940s-70s	Sewage effluent discharged into lagoon.
1954	Lake Sutherland Dam constructed.
1965-67	Interstate 5 constructed.
1986	San Dieguito River Valley Conservancy established.
1989	San Dieguito River Park Joint Powers Authority formed.
2011	San Dieguito Wetlands Restoration Project completed.

Sources: California Department of Public Works 1949, Bronson 1968, Mudie et al. 1976, Marcus 1989, Elwany et al. 1995, Sherman 2001

(top) Undated view of Lake Hodges Dam, constructed 1918. (Ed Fletcher San Diego photograph album, courtesy of Mandeville Special Collections, UC San Diego)

(bottom) Undated photo of a crowd, with band, gathered along the railroad tracks in the community of Del Mar, just south of San Dieguito Lagoon. (photo #79:309, courtesy of San Diego History Center)

STRATFORD INN GARAGE

BATCHELDER & GARDNER
GENERAL MERCHANDISE

San Dieguito Lagoon as shown on the T-sheet, surveyed May to July 1889. (Rodgers and Nelson 1889; courtesy of NOAA)

a The lagoon inlet is depicted as open to the ocean. A faint trace of the coastal road is shown crossing the open inlet.

b The dotted line signifying Mean Lower Low Water implies that when the inlet was open, lagoon channels would have often drained completely at low tide.

c An extensive salt marsh plain occupies the central portion of the lagoon.

d At the time of the survey, the San Dieguito River flowed into the northeastern corner of the lagoon.

e A second sinuous channel lined with woody riparian vegetation enters the estuary at its southern edge, possibly representing an old course of the San Dieguito River.

f A small hill or "island" of raised land, marked by a red contour line, borders the lagoon on the northeastern side, restricting the possible locations of the San Dieguito River's connection to the lagoon.

g Three perennial ponds lie in depressions on the easternmost margin of the lagoon.

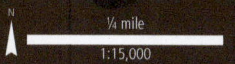

¼ mile
1:15,000

Freshwater / Brackish Wetland

Salt Marsh

Open Water / Mud Flat

Salt Flat (Seasonally Flooded)

Beach

Dune

Stream and Distributary

¼ mile

1:15,000

SAN DIEGUITO LAGOON HISTORICALLY OCCUPIED JUST UNDER 600 ACRES AT THE MOUTH OF THE SAN DIEGUITO RIVER VALLEY. The estuary was dominated by a large salt marsh covering approximately 540 acres and extending well over a mile inland, representing over 90% of the total lagoon area. On its northeastern edge the salt marsh was bounded by a small hill stretching approximately 2,000 feet north to south; its southeastern edge extended east of what is now Interstate 5 before transitioning into perennial ponds nestled in narrow valleys. A beach berm ranging from 100 to nearly 600 feet wide separated the lagoon from the ocean everywhere but at its northern edge. East of the estuary, freshwater and brackish wetlands extended more than two miles inland. Unlike the other five lagoons in the study, no salt flat was documented in San Dieguito Lagoon.

Several sinuous tidal channels connected the San Dieguito River and surrounding valley lands to the ocean, lacing the salt marsh before converging into a single large inlet at the northern edge of the lagoon. At the time the T-sheet was made in the late 1880s, the San Dieguito River entered the lagoon at the northern edge of the valley, flowing into the northernmost tidal channel. A second, willow-lined sinuous channel at the southern end of the valley connected to the southern tidal channel; this may have been either the lowest reach of the creek flowing from La Zanja Canyon (USGS 1903) or possibly the remnants of an old route or channel segment of the San Dieguito River itself (Freeman 1854b, Post 1913).

SALT MARSH: In contrast to other lagoons where salt flat represented the majority of lagoon area, San Dieguito Lagoon was dominated by an extensive pickleweed plain laced with tidal channels (see pages 122-123). One early 20th century image of the marsh is labeled "the meadows" (South Coast Land Co. 1912), a characterization echoed by another contemporary observer who called the marsh the "San Dieguito meadows" and poetically described the "dream-kissed vale of San Dieguito, serpentined with natural canoe-ways that have crept in from the great waters" (McGroarty in Murphy 1915).

SMALL HILL: This apparent hole in the historical synthesis mapping marks the location of an elevated tableland, only about 20 feet high. One 19th century surveyor called the feature a "small island" (Goldsworthy 1874a), reflecting its position as an upland area entirely surrounded by salt marsh and freshwater/brackish wetlands. The hill created an abrupt physical barrier along this section of the lagoon, constricting the location of the San Dieguito River to a narrow opening between the hill and the northern bluffs (or at times possibly diverting flow to the south).

SAN DIEGUITO RIVER MOUTH: The San Dieguito River's larger watershed and flow volume allowed it to be frequently connected to the ocean during the 19th century. However, the mouth still closed periodically (see pages 180-183).

BEACH AND DUNE: A broad (~100-600 feet wide) beach barrier separated San Dieguito Lagoon from the ocean. It was described as less pronounced and sandier (i.e., with fewer cobbles) than the natural berm at Los Peñasquitos Lagoon (Rodgers 1889).

NEW TIDAL CHANNEL: Early 20th century sources show a large channel cutting across the marsh in the middle of the lagoon (e.g., Unknown ca. 1915a, at right and on next page; San Diego County 1928). The channel is not shown on the T-sheet or other 19th century sources, suggesting that flows were first diverted through this area around the turn of the century. Today this channel represents the primary course of the San Dieguito River.

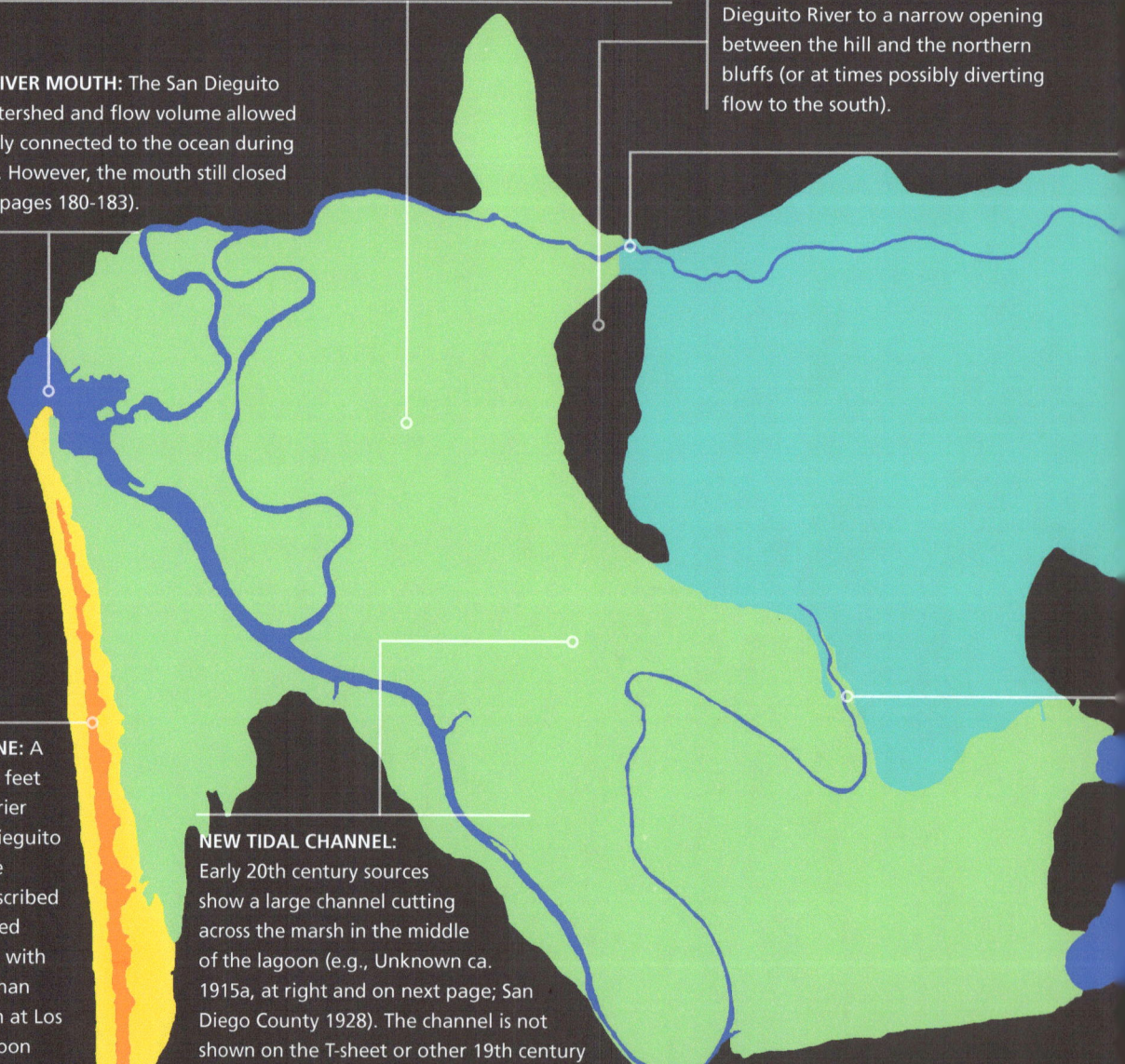

SAN DIEGUITO RIVER AND VALLEY: The San Dieguito River passed through several small valleys before reaching the estuary. The lower river channel was recorded to be about 60-100 feet wide in the valley just above the estuary, with 4-5 foot high banks (Goldsworthy 1874a, Post 1913). Its large watershed and significant wet-season discharge made the river more comparable to other relatively large streams in the region – such as the Santa Margarita, San Luis Rey, and even San Diego rivers – than to the smaller creeks emptying into other lagoons in the study area (see pages 124-125).

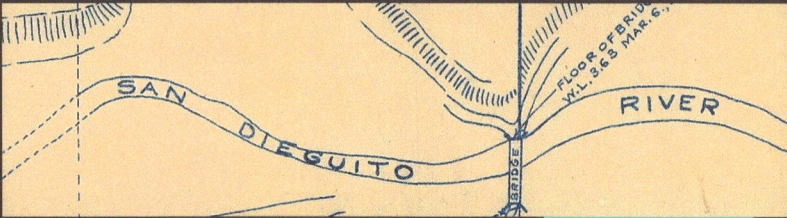

■	Freshwater / Brackish Wetland
■	Salt Marsh
■	Open Water / Mud Flat
■	Salt Flat (Seasonally Flooded)
■	Beach
■	Dune
■	Stream and Distributary

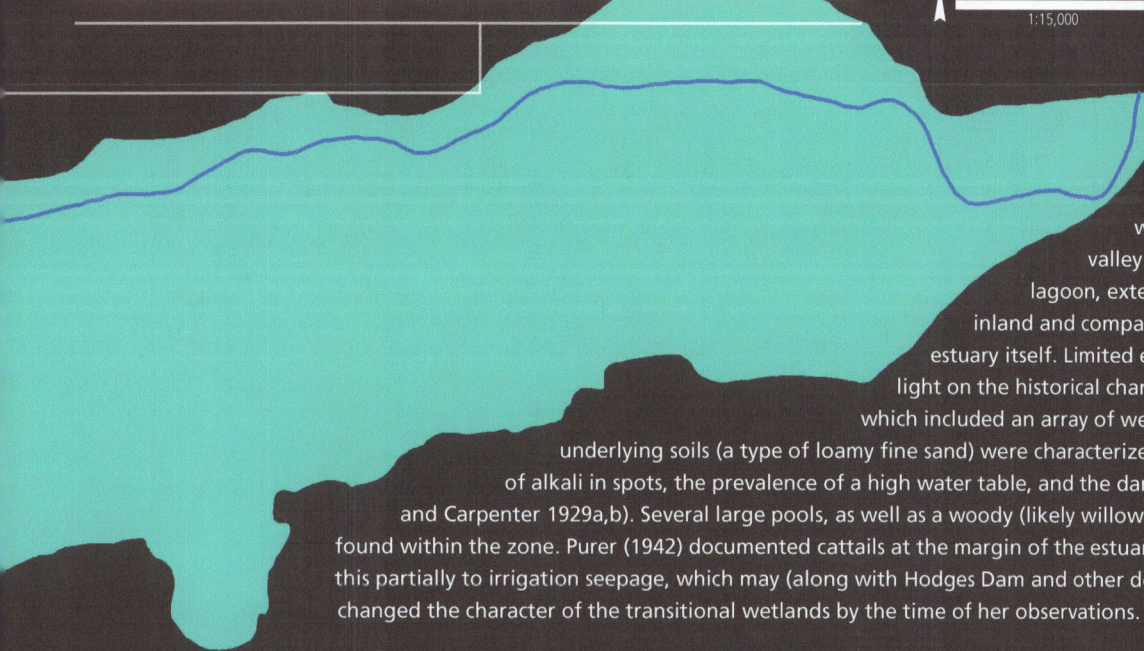

¼ mile
1:15,000

**FRESHWATER/
BRACKISH WETLANDS:**
Freshwater and brackish wetlands covered the valley floor upstream of the lagoon, extending over two miles inland and comparable in area to the estuary itself. Limited evidence sheds some light on the historical character of this zone, which included an array of wetland habitat types. The underlying soils (a type of loamy fine sand) were characterized by an "accumulation of alkali in spots, the prevalence of a high water table, and the danger of overflow" (Storie and Carpenter 1929a,b). Several large pools, as well as a woody (likely willow) riparian corridor were found within the zone. Purer (1942) documented cattails at the margin of the estuary, though she attributes this partially to irrigation seepage, which may (along with Hodges Dam and other developments) have changed the character of the transitional wetlands by the time of her observations.

SERPENTINE CHANNEL: At the time the T-sheet was surveyed in 1889, this southern tidal slough connected to a winding fluvial channel marked by a broad corridor of woody riparian vegetation (likely willows) running along the southern edge of the valley parallel to the main San Dieguito River channel (see T-sheet on pages 116-117). This feature was called "the Lagoon" or "Serpentine" in the early 20th century, and was posited to mark the location of an old course of the San Dieguito mainstem (Post 1913). This is tentatively supported by early survey records, which document crossing the "San Digito River" or "San Bernarda River" [sic] at the southern edge of the valley, near the course shown on the T-sheet and backed up against the small hill (Freeman 1854b, Goldsworthy 1874a). However, Goldsworthy (1874a) also records crossing the river at the 1889 location to the north, so the story is far from clear. It appears that the river has occupied numerous channels in the lower valley in recent history; further research may help clarify the sequence of shifts in channel location upstream of the estuary. Remnants of this willow corridor persisted at least until the mid-1930s (Knox 1933-4).

Salt Marsh: Unknown ca. 1912 in South Coast Land Co. 1912, courtesy of The Bancroft Library; Tidal Channel: Unknown ca. 1915a, courtesy of San Diego History Center; River and Valley: South Coast Land Co. 1913, courtesy of Holdings of Special Collections & Archives, UC Riverside; Serpentine Channel: Rodgers and Nelson 1889.

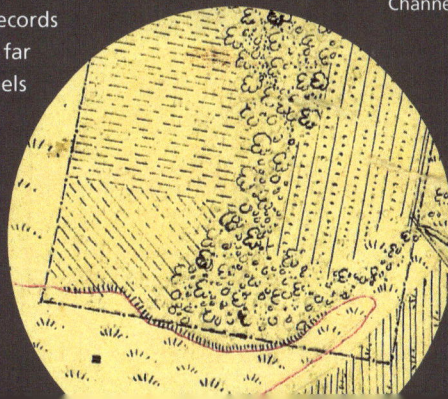

Panorama of San Dieguito Lagoon showing uniform marsh plain and sinuous tidal channels, ca. 1915. This image is a composite of two photographs and looks north across the estuary; the ocean is visible at far left. The large channel cutting across the marsh plain at right (detail shown on previous page) was not present at

the time of the T-sheet survey in the 1880s, though today it represents the primary course of the San Dieguito River. The structure in the foreground is a sugar beet unloading facility. (Unknown ca. 1915a, courtesy of San Diego History Center)

SAN DIEGUITO LAGOON: San Dieguito River and Valley

On Saturday, July 15, 1769, the Portolá Expedition entered San Dieguito Valley (which they called *San Jácome de la Marca)* from the south, east up the valley from the lagoon. Members of the expedition found the valley very attractive and verdant for mid-summer, so much so that they thought it would make a good location to establish a mission: "entirely covered with pasture, with some groves of trees, and has much water collected in pools" (Costansó and Browning 1992). They also noted a "very lush stream" with many pools running on the northern side of the valley: the San Dieguito River (Crespí and Brown 2001).

Along with the Santa Margarita, San Luis Rey, and San Diego rivers, the San Dieguito River was considered to be one of the major streams in the region, draining a much larger watershed than other streams in the study area. During floods, high discharges could flood the lower valley and marsh and open the river mouth (see bottom photo on facing page). Many early travelers noted the difficulty they experienced in crossing the San Dieguito River during the wet season, and especially during floods (e.g., Bell 1869, Duhaut-Cilly [1827]1997, Sturgis 1884 in Ewing 1988, de Anza and Bolton 1930).

As treacherous as the San Dieguito River could be in winter, historical records show that the river's dry-season flow was quite limited in many reaches, particularly where the river passed through broad alluvial areas such as the San Pasqual and San Bernardo valleys (see top photo on facing page). Numerous 19th century sources report intermittent surface flow in these reaches, which limited early large-scale efforts to irrigate these valleys:

> These [irrigation] endeavors…are lacking in importance on account of the intermittent nature of the stream, whose waters in summer are seldom visible above the surface of the sand that fills its bed throughout in all the valleys below the Santa Ysabel. (Hall 1888)

> The valley of San Pascual is six miles in length, and the river here has so slight a fall, and such a broad, sandy bed, that its summer flow is wholly lost and absorbed. (Hall 1888)

> In summer it is generally dry throughout its whole course, from above the Indian village of San Pascual; but in winter it is a large stream. (W.W. Elliott & Co. 1883)

> Above San Pasqual Valley the creek [San Dieguito River] maintains a light flow throughout the year, but below that point the channel is dry during the summer months. (Freeman et al. 1912)

These qualitative descriptions are supported by the limited early 20th century gage data (pre-dating Hodges Dam) available for the river at San Bernardo valley (USGS gage data 1912-1915, #11029500), which document extended periods with no discharge in summer and fall.

In the river's lowest reaches (e.g., through the mapped freshwater/brackish wetlands), accounts indicate the presence of perennial water "running sluggishly to the sea" (Hall 1888; see also Post 1913).

(top) "View in San Pasqual Valley—Bernardo [San Dieguito] River." The season in this undated (ca. 1887) photograph is unknown, but it shows shallow flow in a reach noted by early sources as having intermittent flow. (Gunn 1887, courtesy of Society of California Pioneers)

(left) San Dieguito River mouth, looking east, during February 1927 floods. The railroad bridge is in the mid-ground; the Highway 101 bridge in foreground. Note flood damage to Highway 101 bridge. (Unknown 1927, courtesy of San Diego History Center)

SAN DIEGUITO LAGOON: Change Over Time

MUCH OF THE HISTORICAL WETLAND FOOTPRINT AT SAN DIEGUITO LAGOON HAS BEEN CONVERTED TO DEVELOPED AREAS, with a few exceptions (e.g., portions of the tidal channel network, some salt marsh, and several small ponds on the eastern margin of the lagoon). Substantial salt marsh area has been lost, particularly on the northern side of the lagoon where the Del Mar Fairgrounds is located today. In addition, there has been widespread loss of historical freshwater/brackish wetlands upslope of the lagoon; large portions of this area are now occupied by a golf course and an equestrian complex.

Within the past decade, however, large areas of salt marsh have also been created in formerly transitional habitats east of the historical lagoon as part of the San Dieguito Wetlands Restoration Project (SDG&E 2013). The project also involved the construction of berms, the dredging of the river mouth and tidal channels, and the creation of new subtidal basins. (The contemporary wetland mapping used below was developed prior to the completion of the restoration project, and thus does not reflect some of the changes visible in more recent aerial photographs.)

One of the most notable changes since the 19th century has been a shift in where the San Dieguito River enters the estuary. In 1889, the main river channel ran along the northern edge of the valley, connecting to northern branches of the tidal slough network (Rodgers and Nelson 1889). Sources from as late as the 1910s continue to show the river in this location (USGS 1903, South Coast Land Co. 1913). By the 1920s, however, the main fluvial connection had shifted to the south, connecting to a broad (~100-200 foot wide) channel not shown on the T-sheet (San Diego County 1928, Knox 1933-4; see middle image on facing page). While a small channel is still visible to the north in sources from this period (apparently occupying the formerly dominant northern route), it was clearly secondary by this time. The southern channel shown by early 20th century sources is approximately the same course followed by the river today.

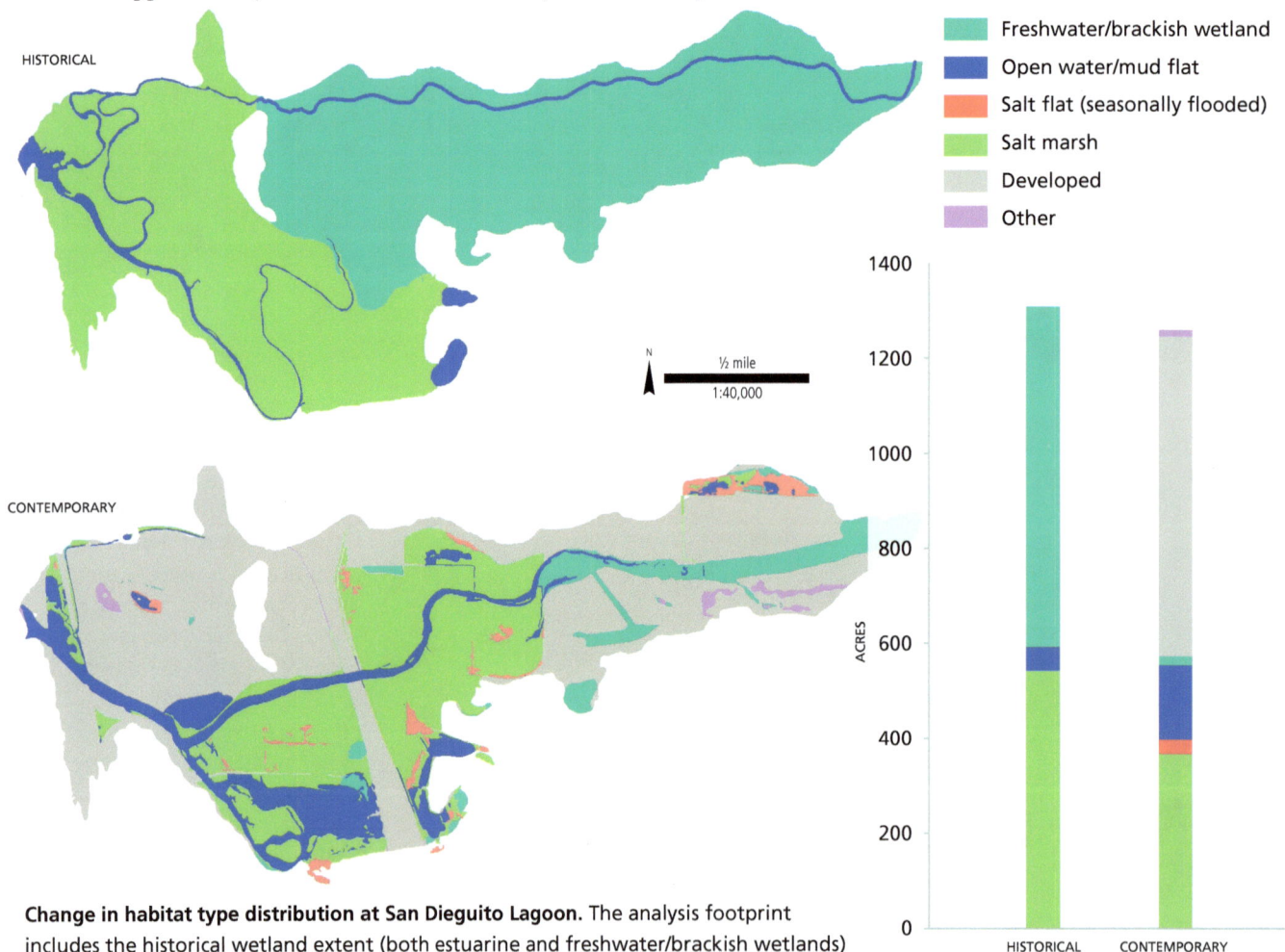

Change in habitat type distribution at San Dieguito Lagoon. The analysis footprint includes the historical wetland extent (both estuarine and freshwater/brackish wetlands) as well as additional contemporary estuarine areas.

1889

1928

2009

½ mile

1:40,000

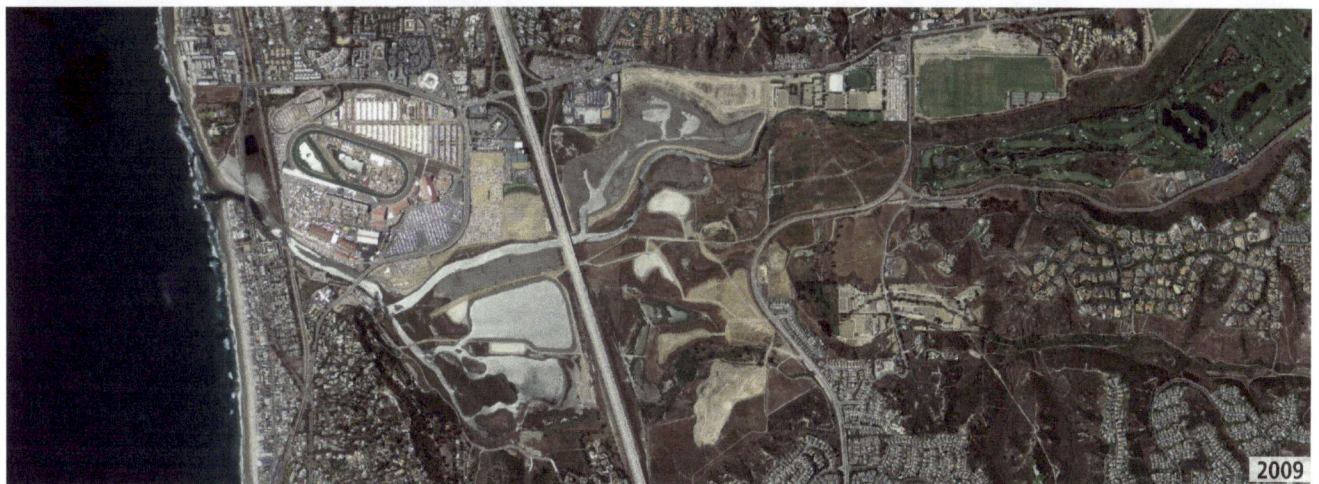

San Dieguito Lagoon was historically dominated by salt marsh, with extensive freshwater/brackish transitional wetlands on the inland side of the lagoon extending up the valley **(top)**. As can be seen in the 1928 aerial photos **(middle)**, substantial changes had occurred within the lagoon by the early 20th century, including degradation of the salt marsh and alterations in the configuration of the San Dieguito River. The contemporary estuary **(bottom)** bears little resemblance to the historical lagoon, though restoration efforts in the past decade have significantly increased the extent of salt marsh within the wetland complex. (Rodgers and Nelson 1889, San Diego County 1928, NAIP 2009)

(top) This ca. 1954 photo, looking east across San Dieguito Lagoon and up the river valley, shows many of the major changes that had occurred by the mid-20th century. Del Mar Fairgrounds occupies a large portion of the historical salt marsh area on the left side of the image. The San Dieguito River, depicted in early sources on the northern side of the valley, can be seen flowing around the Fairgrounds to the south. Highway 101 and the railroad berm cross the river in the foreground. (Collection 87-26, courtesy of Scripps Institution of Oceanography Archives, UC San Diego)

(right) Remnant ponds on the eastern margin of the historical lagoon area. Topographically confined by uplands on three sides, these ponds (shown as perennial open water features by the 1889 T-sheet) continue to support open water for parts of the year. A distinct transition between salt marsh and freshwater/brackish vegetation is visible in the left image. (photos by Sean Baumgarten, January 2013)

9. LOS PEÑASQUITOS LAGOON

On the morning of November 20, 1886, stepping off the cars at Del Mar, we spent a little time in walking upon the beach…A short distance to the south, the route passes down into a narrow canyon running parallel with the coast…Not far to the east of this, the Quade and Carroll canyons, having united in mid-mesa, make out into the Soledad, and the Soledad empties into the Peñasquitos, the Peñasquitos into the Cordero Delta and the Cordero Delta into the ocean between the Sorrento hills and the heights of Del Mar.

—ALBERT MATSON 1889, IN EWING 1988

Los Peñasquitos Lagoon (photo by Sean Baumgarten, January 2013)

Los Peñasquitos Lagoon is located on the northwestern side of the City of San Diego, just south of the city of Del Mar. The lagoon's approximately 500 acres of wetlands are included within Torrey Pines State Natural Reserve. The lagoon is crossed by several transportation corridors, including the Santa Fe Railroad line (which cuts through the center of the lagoon) and Highway 101 (which runs along its western edge). Interstate 5 lies just to the east of the lagoon.

Los Peñasquitos Lagoon is fed by three streams: Carroll Creek, Los Peñasquitos Creek, and Carmel Creek, which cumulatively drain an approximately 95 square mile area. Carmel Creek drains into the lagoon from the north through Carmel Valley, while Los Peñasquitos and Carroll creeks run through their respective canyons before merging in Sorrento Valley and flowing into the lagoon from the south. Los Peñasquitos Creek drains the largest area, accounting for almost 70% of the total area draining into the lagoon (Crooks et al. 2012). Rapid urbanization in the surrounding watershed beginning in the second half of the 20th century has resulted in increased sediment and water delivery to the lagoon, including increased dry-season flows (Greer and Stow 2003, White and Greer 2006).

Los Peñasquitos ("little cliffs" or "crags" in Spanish) Lagoon has been alternately known as Cordero Slough or Delta, Soledad Lagoon, Torrey Pines Lagoon, or Sorrento Lagoon.

Before the war between the United States and Mexico, the inhabitants of the Pueblo of San Diego were accustomed to make their farms in [La Soledad Valley], it having running water sufficient for corn, beans and other crops; they lived in town but in April and May went out to La Soledad, in order to plant, returning after the harvest; they did not build their permanent houses, using it only for their fields, there being no other land within the Pueblo so suitable for farming.

—JOSÉ ANTONIO SERRANO IN HOWARD 1869

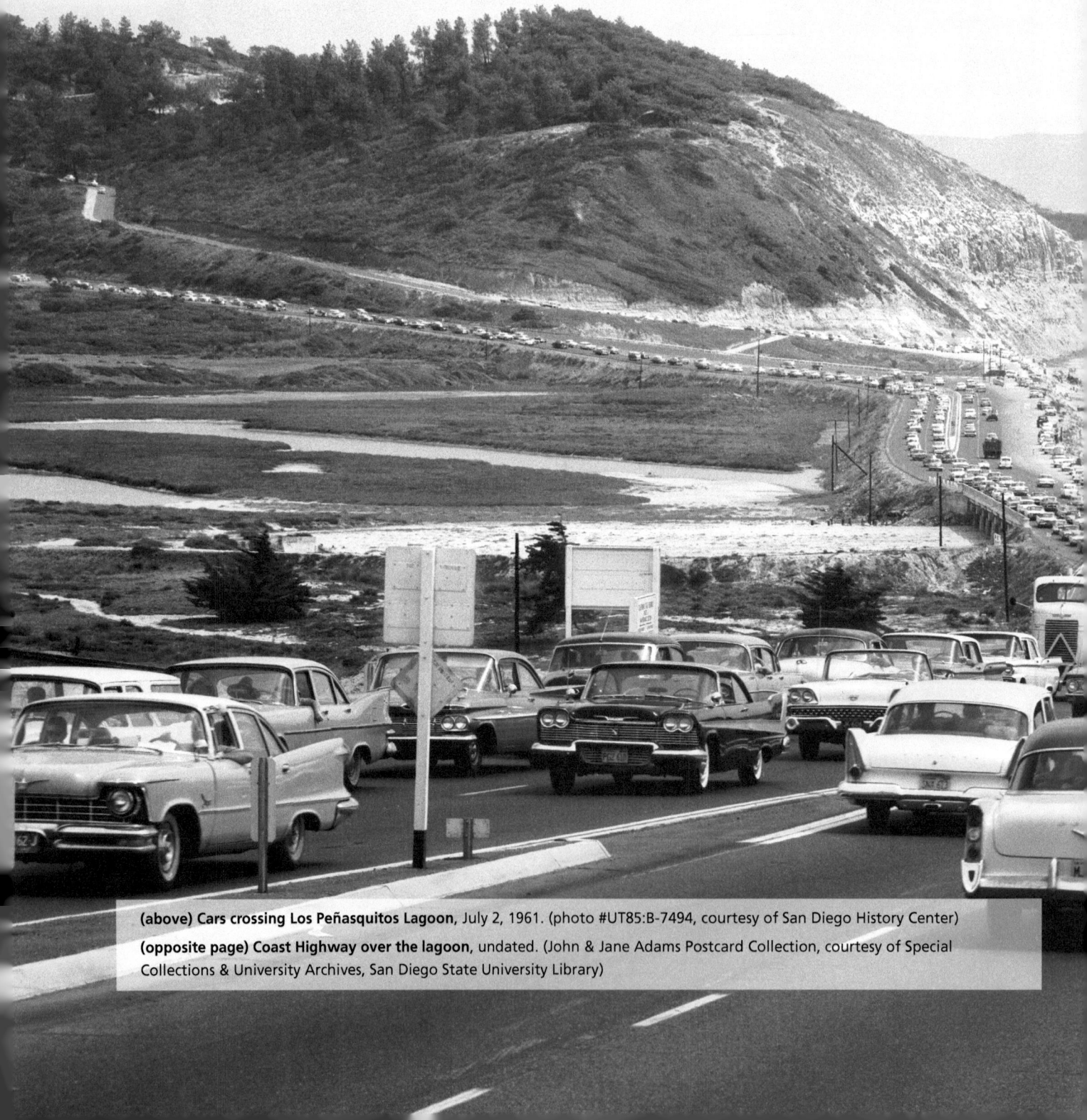

(above) Cars crossing Los Peñasquitos Lagoon, July 2, 1961. (photo #UT85:B-7494, courtesy of San Diego History Center)
(opposite page) Coast Highway over the lagoon, undated. (John & Jane Adams Postcard Collection, courtesy of Special Collections & University Archives, San Diego State University Library)

Timeline: Los Peñasquitos Lagoon

Pre-1848	"La Soledad Valley" [Peñasquitos Valley] used by residents of the Pueblo of San Diego to raise crops.
1823	Rancho Los Peñasquitos granted to Francisco María Ruiz, making it the first land grant in San Diego County.
1881-82	California Southern Railroad line constructed between National City and Oceanside.
1912-15	Pacific Coast Highway constructed.
Early 1900s	Santa Fe Railroad line constructed through middle of lagoon.
1932-33	Highway 101 expanded, including construction of the "Sorrento Overhead" over the Santa Fe Railroad.
1962-72	Sewage effluent discharged into Los Peñasquitos Creek.
1965-67	Interstate 5 constructed.
1968	North Beach parking lot constructed.
1983	Los Peñasquitos Lagoon Foundation established.

R-28 A California Highway

On U. S. Highway 101, near Torrey Pines, Los Angeles to San Diego

Sources: USDC ca. 1869, Wilson 1883, Smythe 1908, Mudie et al. 1974, Greer and Stow 2003, White and Greer 2006, San Diego Water Board 2011

Los Peñasquitos Lagoon as shown on the T-sheet, surveyed May to July 1889. (Rodgers and Nelson 1889; courtesy of NOAA)

a The inlet is shown open at the time of the T-sheet survey. As for San Dieguito lagoon, the location of the Mean Lower Low Water line (dotted, just offshore) indicates that the channels lacing the marsh would have often drained completely at low tide.

b The "Beach Shingle R.R." spur of the California Southern Railroad was constructed in order to transport cobble from the beach and dune complex to pave streets in San Diego (Rodgers 1889).

c A large salt marsh plain is shown as the predominant lagoon feature. No salt flat is depicted, though other early sources – including a survey contemporary with the T-sheet (Unknown 1888a) – do document the historical presence of salt flat.

d Numerous roads are shown crisscrossing the marsh, suggesting that the marsh plain was dry enough at certain times of the year to permit passage.

e The original California Southern Railroad line ran along the eastern side of the lagoon, much further inland than the coastal route taken across the other North County lagoons. As a result, in contrast to other systems the railroad berm did not constrict the lagoon mouth or restrict circulation in large portions of the marsh.

¼ mile
1:15,000

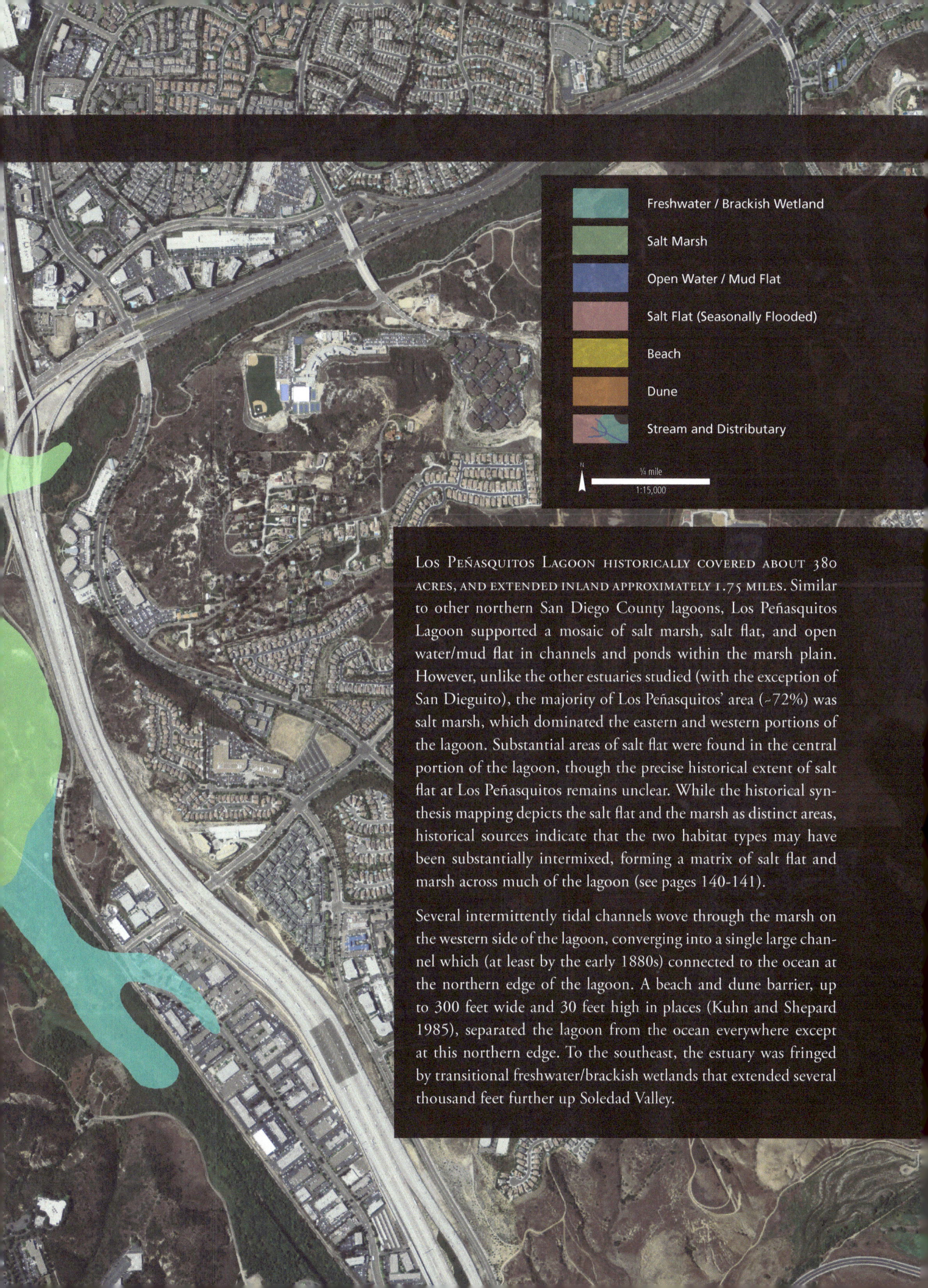

Freshwater / Brackish Wetland

Salt Marsh

Open Water / Mud Flat

Salt Flat (Seasonally Flooded)

Beach

Dune

Stream and Distributary

¼ mile
1:15,000

Los Peñasquitos Lagoon historically covered about 380 acres, and extended inland approximately 1.75 miles. Similar to other northern San Diego County lagoons, Los Peñasquitos Lagoon supported a mosaic of salt marsh, salt flat, and open water/mud flat in channels and ponds within the marsh plain. However, unlike the other estuaries studied (with the exception of San Dieguito), the majority of Los Peñasquitos' area (~72%) was salt marsh, which dominated the eastern and western portions of the lagoon. Substantial areas of salt flat were found in the central portion of the lagoon, though the precise historical extent of salt flat at Los Peñasquitos remains unclear. While the historical synthesis mapping depicts the salt flat and the marsh as distinct areas, historical sources indicate that the two habitat types may have been substantially intermixed, forming a matrix of salt flat and marsh across much of the lagoon (see pages 140-141).

Several intermittently tidal channels wove through the marsh on the western side of the lagoon, converging into a single large channel which (at least by the early 1880s) connected to the ocean at the northern edge of the lagoon. A beach and dune barrier, up to 300 feet wide and 30 feet high in places (Kuhn and Shepard 1985), separated the lagoon from the ocean everywhere except at this northern edge. To the southeast, the estuary was fringed by transitional freshwater/brackish wetlands that extended several thousand feet further up Soledad Valley.

LOS PEÑASQUITOS LAGOON: Points of Interest

LOS PEÑASQUITOS LAGOON INLET: Unlike other lagoons in the region, the Los Peñasquitos inlet was not constricted by the 1880s California Southern Railroad berm, which was constructed much further to the east along the margin of the estuary rather than near the mouth. As a result, the impact of this structure on tidal circulation and mouth location was likely much smaller than for neighboring systems (see also Mudie et al. 1974). Historical sources document that the lagoon's inlet opened and closed intermittently (see table 11.2 on pages 180-181); Wilson (1883) wrote that the lagoon was open during "spring tides and storms." The photo at left shows the lagoon inlet open at low tide in January, 1974.

BEACH AND DUNE: The beach and dune barrier, notable for the prevalence of cobbles, was described as a "high wall composed of small rocks" (Wilson 1883). These cobbles were used in the 19th century for street paving in San Diego; a railroad spur (called the "Beach Shingle Railroad" or the "Sea Wall Spur") was built directly to the beach to transport them (Rodgers 1889).

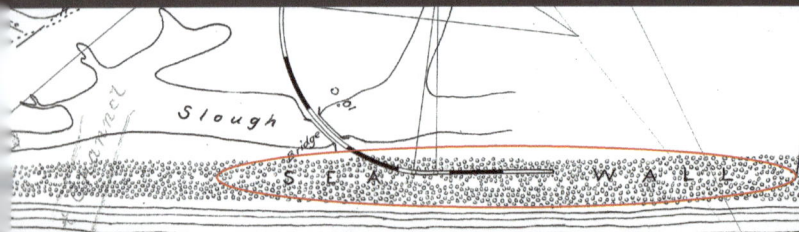

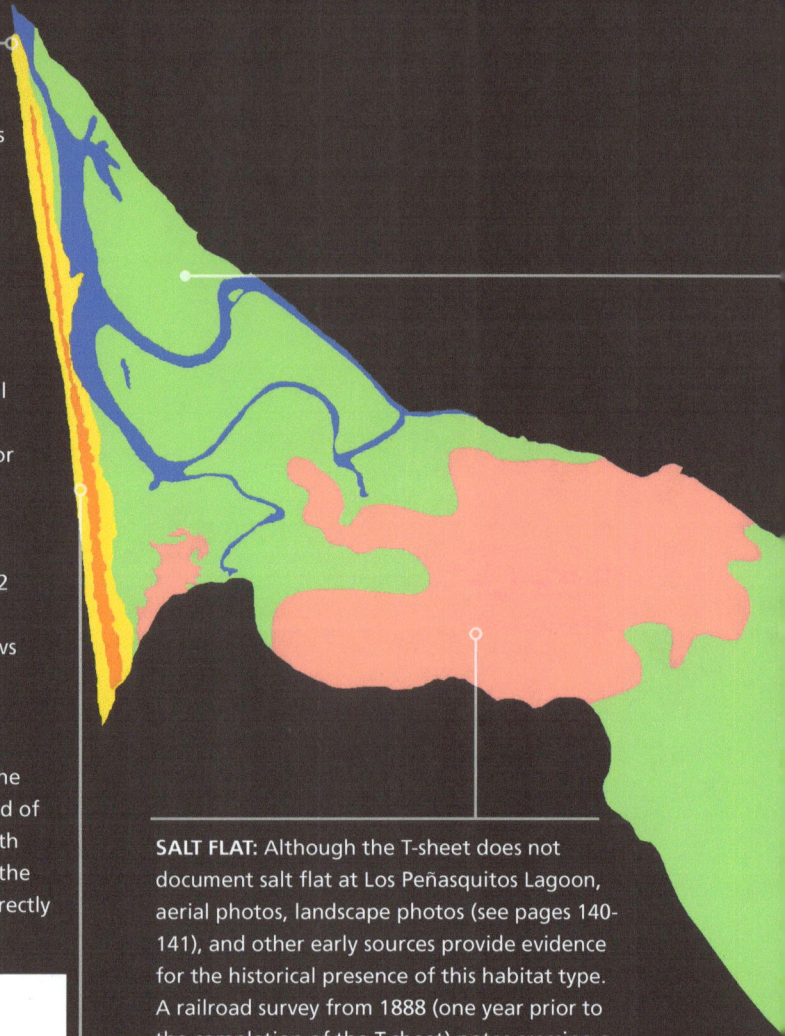

SALT FLAT: Although the T-sheet does not document salt flat at Los Peñasquitos Lagoon, aerial photos, landscape photos (see pages 140-141), and other early sources provide evidence for the historical presence of this habitat type. A railroad survey from 1888 (one year prior to the completion of the T-sheet) notes crossing a broad "alkali flat" in Soledad Valley (Unknown 1888a), though the full extent of the alkali flat is not specified.

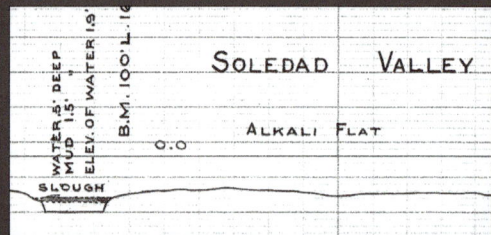

- Freshwater / Brackish Wetland
- Salt Marsh
- Open Water / Mud Flat
- Salt Flat (Seasonally Flooded)
- Beach
- Dune
- Stream and Distributary

N
¼ mile
1:15,000

Los Peñasquitos Lagoon Inlet: USA-C1.9 #74.153-5, courtesy of Scripps Institution of Oceanography Archives, UC San Diego; Beach and Dune Barrier: Unknown 1888b, courtesy of California State Railroad Museum; Salt Flat: Unknown 1888a, courtesy of California State Railroad Museum; Flooding: USA-C1.9 #669-6, courtesy of Scripps Institution of Oceanography Archives, UC San Diego.

3 APR 58   APR · 58 ·   669-6

FLOODING: Large floods could inundate the lagoon, particularly when the mouth was closed. An account of the landmark 1884 floods described the lagoon as "entirely submerged at last accounts" (*San Diego Union* 3/16/1884 in *Los Angeles Herald* 3/20/1884). The above image shows the lagoon similarly flooded during the wet year of 1958.

SALT MARSH: Pickleweed-dominated salt marsh covered approximately 72% of the lagoon area, with large expanses both east and west of the central salt flat (see pages 140-141).

TRANSITIONAL WETLANDS: The Soledad Valley was described as quite lush: in January 1847, it was noted to have "water and a luxuriant thick growth of grass" (Cooke 1849). This apparently persisted into the summer months as well: in mid-July 1769, Crespí described the valley as a "vastly handsome valley or hollow" with abundant wild grapes, wild roses, and large clumps of grass; he noted that the valley was so green and lush that it "seemed nothing other than a field of corn" from the bluffs above (Crespí and Brown 2001). Other early sources support this, documenting the presence of shallow groundwater levels in the lower valley (e.g., Ellis and Lee 1919).

LOS PEÑASQUITOS CREEK FLOW: Los Peñasquitos Creek was the primary tributary into the estuary, draining almost 70% of the lagoon's watershed. Although the valley was lush and water was available near the surface even during the summer, Los Peñasquitos Creek was generally dry through the summer months (as were Carroll and Carmel creeks; USGS 1903). This finding is supported by general statements – one of which, for example, describes the creek as "a small stream which in summer takes refuge underground from the thirsty sun" (Chase 1913) – as well as by two GLO surveys from May 1854 and July 1858, both of which note that the bottom reach of the creek was about 20-30 feet wide and dry (Freeman 1854b, Hays 1858b). Surface water would have persisted only in pools in the dry season: in July 1769 Crespí noted that native residents used a well in the dry bed of the creek for their water supply (Crespí and Brown 2001), and Jepson (1898) recorded pondweed (*Potamogeton* sp.) in "pools in the dry bed of Penasquitas Creek" in June 1897. Even when surface flows stopped, however, groundwater levels were high and continued to provide freshwater input to the lagoon year-round (Benton 1886, Ellis and Lee 1919). The hydrology of Los Peñasquitos Creek and other creeks feeding into the lagoon have been heavily altered by development in the past few decades (see page 144).

Las Peñasquitas is a long, narrow valley threaded by a small stream which in summer takes refuge underground from the thirsty sun.

—CHASE 1913

LOS PEÑASQUITOS CREEK RIPARIAN CORRIDOR: Early descriptions of lower Los Peñasquitos Creek emphasize the presence of riparian sycamores and live oaks (Wilson 1883, Crespí and Brown 2001), a riparian canopy often indicative of intermittent flow conditions. In December 1874, GLO surveyor J. Goldsworthy (1874b) noted a sycamore tree on the bank of Los Peñasquitos Creek a short distance upstream of the lagoon. Chase (1913) observed "scattered sycamores and elders" along the creek channel, and Benton (1886) described "sycamore, live oak and willow." Today, the lower Peñasquitos Creek riparian corridor is instead dominated by dense stands of willows, with some sycamores and oaks flanking on higher ground (White and Greer 2006).

(top) This panorama of the western portion of Los Peñasquitos Lagoon is a composite of three photographs taken from the bluffs to the south of the lagoon between 1911 and 1915. Salt marsh and channels occupy the area of the lagoon closest to the ocean. Further inland, the lagoon is characterized by a large expanse of salt flat with interspersed patches of salt marsh. A railroad berm constructed in the early 1900s runs through the middle of the lagoon. (photos #91:18564-3057, 91:18564-205, 80:6532; courtesy of San Diego History Center)

(left) This 1913 photo, looking southeast, shows the salt flat in the center of the lagoon grading into salt marsh and freshwater/brackish marsh further inland. (photo #91-18564-203, courtesy of San Diego History Center)

A few alkali flats are exposed in summer; but in winter most of the area is covered with water.

—KNOX 1934-5A

LOS PEÑASQUITOS LAGOON: Postcards of the Lagoon

Situated in the valley just north of Torrey Pines State Natural Reserve, Los Peñasquitos Lagoon was a common sight for tourists traveling to the area to see the park and its eponymous pine tree. Many early photographs and postcards capture the western portion of the lagoon, often as a backdrop to the pines from a vantage point on the cliffs just south of the lagoon. This spread shows a sample of these postcards dating from the early 20th century. While some are clearly stylized or even fanciful representations of the lagoon (e.g., opposite page, top), most appear to be quite accurate depictions of lagoon habitats. Notice, for example, the marsh vegetation detail (opposite page, second from top) and the consistent depiction of channel configuration.

e View of Del Mar from Torrey Pines, near San Diego, Cal.

POST CARD

Miss Lucy Crouch

Oceanside,

California

THIS SPACE FOR THE ADDRESS.

in the Distance, San Diego, Cal.

Opposite, second from top: Photograph Collection, CO-San Diego (A-R), Box 076, courtesy of California Historical Society; others: John & Jane Adams Postcard Collection, courtesy of Special Collections & University Archives, San Diego State University Library.

COAST HIGHWAY FROM TORREY PINES PARK,
SAN DIEGO, CALIFORNIA—54

680 COAST HIGHWAY, NEAR DEL

4932-29

"Torrey Pines," La Jolla, Cal.

106128

LOS PEÑASQUITOS LAGOON: Change Over Time

The distribution and extent of habitats composing Los Peñasquitos Lagoon have changed substantially over the past 150 years. Salt marsh, which historically covered 270 acres (~72% of the estuarine area), has decreased considerably, though it still occupies nearly 160 acres. Freshwater/brackish marsh has expanded from roughly 40 acres historically to over 190 acres today, a ~150% increase.

The salt flats that characterized the central portion of the lagoon have disappeared, replaced largely by salt marsh. In the early 1900s the lagoon was dominated by an extensive salt flat area, visible in the 1928 aerial photos (Greer and Stow 2003; see facing page). It is unclear whether the salt flat extent shown in the aerial photos is representative of earlier conditions, or if salt flat area expanded between the late 19th and early 20th centuries. There were still large areas of salt flat in the middle of the lagoon in the 1950s-70s (Bradshaw 1968, Mudie et al. 1974): Mudie et al. (1974) notes 90 acres of salt flat (about 23% of the total lagoon area – almost identical to the historical acreage), with salt marsh (tidal and non-tidal) accounting for a little over 60% of the total area. Today, however, salt flat is limited to a few small patches totaling less than five acres.

The salt marsh area historically found in the eastern portion of the lagoon is now largely occupied by freshwater/brackish wetlands. The almost complete conversion from salt marsh to freshwater/brackish marsh in this part of the lagoon appears to have been driven largely by sewage discharge in the 1960s and 70s, and later by increased runoff (especially during the dry season) and sedimentation associated with rapid urbanization (Nordby and Zedler 1991, Greer and Stow 2003, White and Greer 2006).

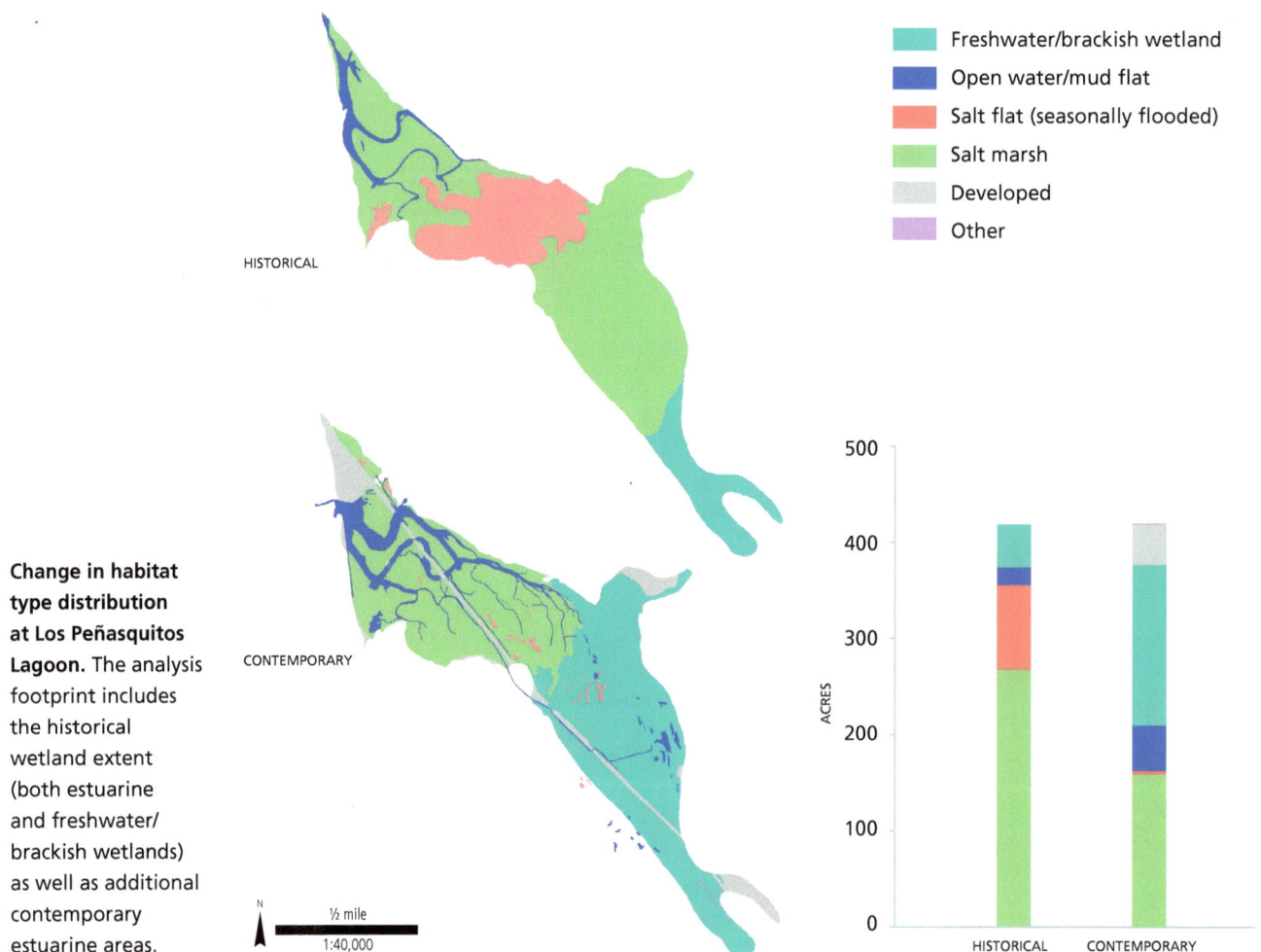

Legend:
- Freshwater/brackish wetland
- Open water/mud flat
- Salt flat (seasonally flooded)
- Salt marsh
- Developed
- Other

HISTORICAL

CONTEMPORARY

Change in habitat type distribution at Los Peñasquitos Lagoon. The analysis footprint includes the historical wetland extent (both estuarine and freshwater/ brackish wetlands) as well as additional contemporary estuarine areas.

½ mile
1:40,000

1889

1928

2009

N
½ mile
1:40,000

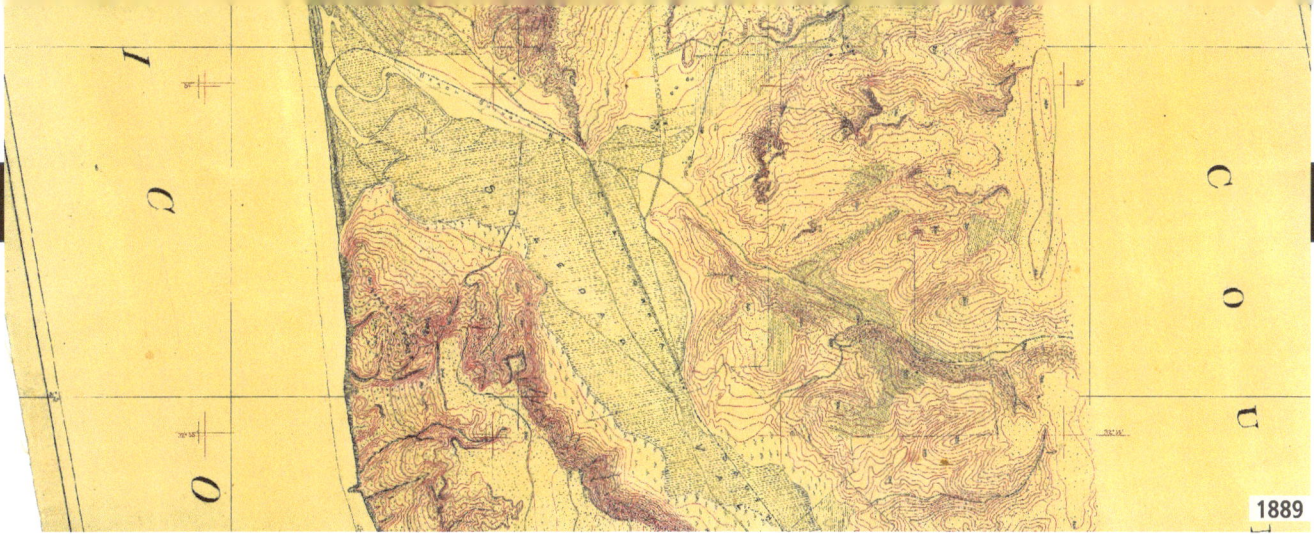

Though the 1889 T-sheet **(top)** shows Los Peñasquitos Lagoon dominated by salt marsh, the lagoon historically also supported substantial salt flat acreage. The 1928 aerial photos **(middle)** show that salt flat was a dominant habitat type in the early 20th century, perhaps occupying an even greater area than in the mid- to late- 19th century. Significant areas of salt flat persisted until the 1970s, but were largely replaced by salt marsh during the later part of the 20th century **(bottom)**. In the eastern lagoon, most of the historical salt marsh has been converted to freshwater/brackish marsh, likely as a result of increased freshwater and sediment inputs from sewage discharge and urban runoff. (Rodgers and Nelson 1889, San Diego County 1928, NAIP 2009)

10. REGIONAL SYNTHESIS:
ECOLOGICAL PATTERNS AND CHANGE

*T*he previous six chapters discussed study results for each lagoon, with a focus on historical ecological patterns and trajectories. The next two chapters take a broader view, synthesizing these findings to provide a regional perspective on historical ecological and hydrologic patterns and physical processes across the study area. In this chapter (Chapter 10) we examine regional ecological patterns, reviewing both similarities and differences in habitat type distribution across the lagoons and discussing habitat type structure, composition, and formation. We also evaluate the ecological functions historically provided by the lagoons and assess changes in lagoon habitat type distribution over time. In the subsequent chapter (Chapter 11) we take a closer look at some of the key physical processes that influenced the historical ecological patterns documented for the lagoons, including freshwater flows, sediment inputs, and inlet dynamics.

Regional Ecological Patterns

A diverse array of habitat types were historically represented across the six northern San Diego County lagoons. Salt flat (about 1,230 acres) and salt marsh (1,330 acres) constituted the majority of the area of these systems; open water and seasonally intertidal mud flat (140 acres) composed the rest of the estuarine area. Beach and dune complexes (120 acres) separated the lagoons from the ocean, and freshwater/brackish wetlands (1,650 acres) were found at the back edge of each estuary, creating a gradual ecotone. The graph on page 165 summarizes the total mapped acreage of each habitat type.

Comparison of the extent and distribution of each habitat type across these six systems reveals a number of ecological patterns at the regional scale. All systems were dominated by habitats types relatively high in the tidal frame (e.g., salt marsh and seasonally flooded salt flat), which together composed 95% of the estuarine area (i.e., excluding upslope transitional freshwater/brackish wetlands) across these systems. Salt flats were found in nearly every system, often composing over half of the total estuarine area. In contrast, mud flat and perennial open water constituted a relatively small percentage (5%) of estuarine area: no lagoon was documented to sustain significant areas of subtidal or perennial deep water outside of small ponds and channels in the marsh plain, and channel networks within the marsh

(top) **Waterbirds in San Elijo Lagoon, January 2013.** (photo by Sean Baumgarten)

appear to have been relatively undeveloped. All lagoons also supported freshwater/brackish wetlands extending inland from their eastern edge, often for many miles.

Despite these similarities, no two lagoons were alike, and the relative proportions of each component varied from estuary to estuary (see figure below). The proportion of salt flat varied from none (San Dieguito Lagoon) to nearly 85% of total estuarine area (Batiquitos Lagoon). Some systems were salt marsh dominant (Los Peñasquitos, San Dieguito), while others were salt flat dominant (Buena Vista, Agua Hedionda, Batiquitos, San Elijo). Differences were also documented in total estuarine area (ranging from approximately 310 acres at Buena Vista Lagoon to nearly 600 acres at San Dieguito Lagoon) and the degree of development of the tidal channel network. While the drivers behind these differences have not been exhaustively studied, they may be at least partly attributable to differences in watershed area and differences in the position of the estuaries within the littoral cell. For instance, lagoons that had the highest proportion of salt flat relative to other habitat types have the smallest watersheds, while San Dieguito Lagoon, whose watershed is about ten times the area of Buena Vista Lagoon's, supported no salt flat. These differences may have implications for sediment supply, freshwater input, inlet dynamics, and tidal circulation that would have affected the ecological characteristics of the lagoons.

The following sections provide more detail on salt marsh and salt flat, the two historically dominant estuarine habitat types within the lagoon complexes, as well as freshwater/brackish wetland (for descriptions of all of the habitat type classifications used in the historical synthesis mapping, see page 30). The discussion draws upon both local historical data and contemporary research from comparable systems.

Historical habitat type distribution by lagoon. Lagoons exhibited regional similarities (such as the prevalence of salt marsh/salt flat and the presence of upslope freshwater/brackish wetland complexes) as well as variations in the extent of each habitat type.

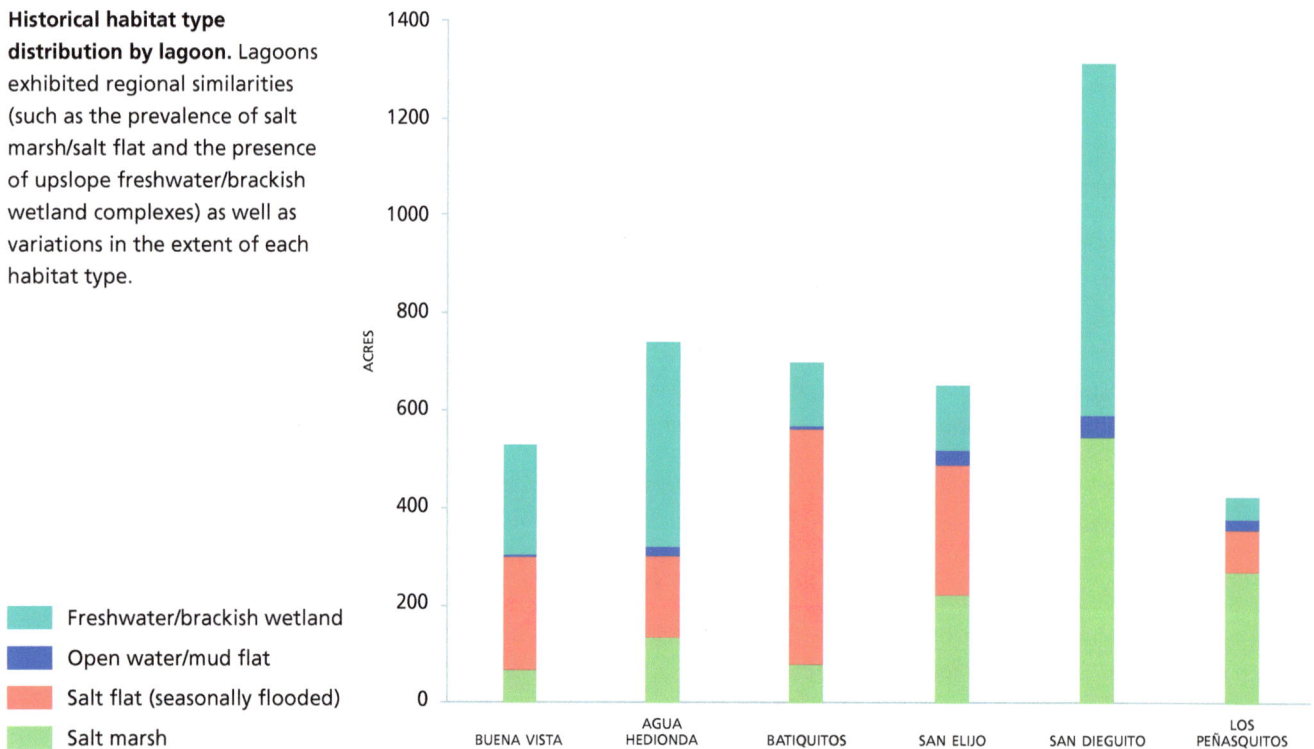

- Freshwater/brackish wetland
- Open water/mud flat
- Salt flat (seasonally flooded)
- Salt marsh

Salt Marsh Plant Communities

In many estuaries, salt marsh vegetation is divided into zones based on periodicity of flooding and levels of salinity (Purer 1942, Hinde 1954, Vogl 1966, Neuenschwander et al. 1979, Callaway et al. 1990). Drawing from a range of salt marshes from central California to Baja California, a brief review of these plant zones is worthwhile because it helps to interpret the plant species found in North County lagoons within a regional context.

In general, the more regular zonation between low, middle, and upper marsh found in fully tidal systems appears to have been absent at most North County lagoons. Though robust early botanical records are lacking, accounts suggest the predominance of *Salicornia* species (e.g., Rodgers 1887-8b, Storie and Carpenter 1929b, Purer 1942). In contrast, records of low marsh species such as *Spartina* are almost completely lacking, consistent with findings from other intermittently closed estuaries in central and southern California (Grewell et al. 2007). Zonation patterns as documented in other estuaries, along with relationships to observations in North County lagoons, are described in more detail below.

LOW MARSH Indicator vascular plant species of low-elevation marshland are California cordgrass *(Spartina foliosa)*, seaside arrowgrass *(Triglochin maritima)*, and widgeongrass *(Ruppia maritima)* (Purer 1942, Vogl 1966, Neuenschwander et al. 1979, Peinado et al. 1994). Each of these is an aquatic plant, growing where it is submerged at least twice a day (or constantly if the system is not tidal) in the lower littoral zone (Hinde 1954). Of these, widgeongrass is a species of fresh and brackish water. For these species in the six lagoons, records only show widgeongrass from Agua Hedionda Lagoon in 1948, and California cordgrass from Los Peñasquitos Lagoon in 1942 (Purer 1942). The only other records are from more recent times, after significant alterations to lagoon hydrology.

A second set of species can be found both in low tidal marsh and in middle marsh, which is inundated by high-high (spring) tides. These include pickleweed *(Sarcocornia pacifica)*, Virginia glasswort *(Salicornia depressa)*, dwarf saltwort *(Salicornia bigelovii)*, and Parish's glasswort *(Arthrocnemum subterminale)*. Although these species can be found in areas inundated regularly by tidal flows, they are not indicators of that condition (Vogl 1966, Neuenschwander et al. 1979, Callaway et al. 1990, Peinado et al. 1994). Of these species, the *Salicornia* species tend to be found in areas with greater flooding, while *Arthrocnemum* is found in areas subject to less frequent flooding (Callaway et al. 1990, Pennings and Callaway 1992). Such flooding, however, may derive from either tidal flow or lagoonal ponding.

MIDDLE MARSH Characteristic species of a middle marsh, which is flooded with spring tides (Hinde 1954), include saltmarsh bulrush *(Bolboschoenus maritimus)*, salt marsh bird's beak *(Chloropyron maritimum)*, saltmarsh dodder *(Cuscuta salina)*, smooth flatsedge *(Cyperus laevigatus)*, and tule *(Schoenoplectus acutus)* (Purer 1942, Vogl 1966, Neuenschwander et al. 1979, Callaway et al. 1990). Historical records show evidence of this community in two of the lagoons studied. In 1942, Buena Vista lagoon had smooth flatsedge and tules, while Los Peñasquitos Lagoon had saltmarsh dodder and smooth flatsedge (Purer 1942). Other indicators of this habitat were recorded much later and following hydrological modification.

A series of species can be seen as transitional indicators into upper marsh, which is flooded only rarely (Purer 1942, Hinde 1954). These can be found both in middle and

Salicornia is the most abundant genus in the lower areas, while *Frankenia* is more frequently found in the upper, slightly raised places of the river valley. *Cressa*, *Heliotropium*, and *Distichlis* are also fairly common, while *Typha* and *Scirpus* have established compact colonies on the edges of the marsh. On the islands which are raised above the salt flat there is mostly *Salicornia*, which grows very luxuriantly, but where it has migrated to the salt flat, its growth is meager... In the marshland above the highway which crosses at its upper end *Atriplex*, *Aplopappus*, and *Cotula* are present. Here there is more seepage of fresh water and the percentage of salinity is low. As one ascends the lagoon area *Cyperus*, *Typha*, and others come in, while *Frankenia* appears where cattle and horses pasture."

—PURER 1942, DESCRIBING VEGETATION AT BUENA VISTA LAGOON

Table 10.1. Selected upper marsh species documented in northern San Diego County lagoons listing earliest data recorded in available herbarium records, 1894-2005. (data from Purer 1942, Consortium of California Herbaria)

	Buena Vista	Agua Hedionda	Batiquitos	San Elijo	San Dieguito	Los Peñasquitos
Acmispon [=Lotus] strigosus				1942		
Amblyopappus pusillus	1942	2004			1901	1942
Anemopsis californica		1936				2004
Aphanisma blitoides					1894	
Cressa truxillensis	1925	2004	2005	1970	1894	1916
Distichlis [=Monanthochloe] littoralis		1942		1942	1942	
Frankenia salina [=Frankenia grandifolia]	1942	1942	1938	1942	1894	2004
Heliotropium curassavicum var. *oculatum*	2004	1938	2005	1942	1970	1942
Jaumea carnosa	2004	1936	2005	1942	1970	1938
Juncus acutus var. *sphaerocarpus*	2004	1936	2005	1942	1970	2005
Limonium californicum		1938		1942	2004	
Spergularia macrotheca					1942	
Spergularia salina					1942	

high marshes in tidal systems and include Watson's saltbush *(Atriplex watsonii)*, saltgrass *(Distichlis spicata)*, and seepweed species *(Suaeda californica, S. taxifolia)*. Purer documents saltgrass at each of the San Diego lagoons except Buena Vista. Agua Hedionda was documented with all three of these species in 1942. Records show *Suaeda taxifolia* at several of the lagoons recently (Agua Hedionda, San Elijo) and historically (San Dieguito; Gander 1936).

UPPER MARSH The upper marsh is only subject to flooding from storm tides in fully tidal systems such as San Francisco Bay (Hinde 1954), but similar salinity and flooding conditions could be created in a seasonally tidal system as well. It occurs above salt flats (where present) in the zonation described for Carpinteria Marsh (Callaway et al. 1990, Pennings and Callaway 1992). The most frequent species in this zone is Parish's glasswort, but given its presence in lower zones, it is not a good indicator. Characteristic species were historically present at each of the northern San Diego County lagoons (table 10.1 above). In addition, the rare Coulter goldfield (salt marsh daisy; *Lasthenia glabrata* ssp. *coulteri*) was documented in the 20th century on the salt flat fringe at both Batiquitos and Los Peñasquitos lagoons (County of San Diego 1979, Williams 1996).

Salt Flats

Salt flats occur in arid and semi-arid landscapes across the world in both inland and coastal environments (Handford 1981), from the Pacific coast of southern California and Baja (Warme 1971, Holser et al. 1981, Callaway et al. 1990) to the Persian Gulf (e.g., Purser 1973, Al-Farraj 2005). They often occur in low-latitude estuaries where evaporation seasonally exceeds inflow, as well as in estuaries with low rainfall, strong seasonal variation in precipitation, and/or irregular tidal inundation (Pennings and Callaway 1992, Largier et al. 1997, Pennings and Bertness 1999). Salt flats are distinct from the small salt pannes that occur in

Nine streams reach the sea between San Mateo Point and La Jolla. The lower parts of all their valleys have broad, flat, marshy bottoms and contain lagoons that on drying up in summer leave broad tracts heavily coated with salt.

—ELLIS AND LEE 1919

areas with limited drainage within many salt marshes (Boston 1983, Pennings and Bertness 2001). They are referred to by a plethora of terms, including salt pans, barrens, alkali flats, salinas, playas, and sabkhas (Briere 2000, Yechieli and Wood 2002).

As the name suggests, salt flats are extremely flat: Callaway et al. (1990), for instance, demonstrated that the salt flat at Carpinteria Salt Marsh was much flatter than any other part of the estuary, with a slope of only about 0.01 inch/foot. In the North County lagoons, this extremely low slope, coupled with the salt flats' position between the marsh plain and the alluvial fan (see diagrams on pages 192-199), produced shallow depressions or basins with limited drainage conditions that trapped water during the rainy season and then concentrated salts as water evaporated during the dry season (Warme 1971, Callaway et al. 1990).

The presence of salt flats at nearly all of the northern San Diego County lagoons is consistent with the process of salt flat formation and the resulting geographic patterns observed in other systems (Pennings and Bertness 2001). Conditions that contribute to salt flat development vary on a north to south gradient, with higher thermal stress resulting in the development of more salt flats in southern California and the southern Atlantic Coast than farther north on either coast (Pennings and Bertness 2001). In more southerly marshes, salinity initially increases with distance from the sea, reaching a maximum that may coincide with the presence of a salt flat before decreasing farther upstream. By contrast, more northerly marshes tend to decrease in salinity from the ocean inland (Pennings and Bertness 2001).

North County lagoons experienced large seasonal fluctuations in salinity as a result of the strongly seasonal rainfall patterns typical of Mediterranean climates. During the dry season, high evaporation rates and low levels of freshwater input and tidal flushing created hypersaline conditions, with soil and water salinities often in excess of 40 parts per thousand (Purer 1942, Carpelan 1969, Dailey et al. 1974, Bradshaw and Mudie 1972, Marcus 1989). Salt flats are the extreme expression of hypersaline conditions, supporting soil salinities often exceeding 100-200 ppt (Purer 1942, Day 1981, Pennings and Bertness 1999, Pennings and Bertness 2001). In the wet season, however, greater freshwater inflows resulted in lower salinities (Purer 1942). This variability is characteristic of many Mediterranean-climate and southern California estuaries (Pennings and Callaway 1992, Largier et al. 1997).

In some systems salt flats occur at elevations where tidal inundation is rare (e.g., Phleger and Ewing 1962, Pierre et al. 1984), allowing salt flats to form even when the estuary is consistently connected to the tides. In Carpinteria Salt Marsh, for example, the salt flat was documented between Mean Higher High Water and maximum high water (Callaway et al. 1990). In northern San Diego County, however, evidence suggests that salt flats occurred lower in the tidal frame, slightly above Mean High Water (Rodgers 1887-8b), and accumulated salt during dry periods when lagoons were cut off from tidal influence. Because of the salt flats' position between marsh plains and alluvial fans, salt flat sediments are fine grained – Reineck and Singh (1973) describe "clayey silt" as the general salt flat substrate, and locally T-sheet surveyor Rodgers (1887-8a) noted "black loam or A-do-be" beneath the crust of salt.

The salt flat inundation regime in North County lagoons was shaped by complex interactions between seasonal fluvial flooding and diurnal tidal fluctuation. When a lagoon was closed during the winter, its salt flats shallowly flooded from precipitation, runoff, and fluvial flows until the mouth breached; 20th century observations recorded depths ranging from a few inches to more than four feet (Mudie et al. 1974, Mudie et al. 1976, Meyer 1980). Once open to tidal circulation, a lagoon's salt flat was intermittently inundated by the tides. At Carpinteria Salt Marsh, where the most relevant modern studies of salt flat development and plant zonation in southern California have been conducted, Pennings and Callaway

FROM GLITTERING *SALINAS* TO BARREN SALT FLATS: CHANGING PERCEPTIONS OF SALT FLATS

On July 9, 1847, Robert Bliss (a private in the Mormon Batallion during the Mexican-American war) spotted one of the northern San Diego County estuarine salt flats from afar. He recorded what he saw in his journal:

> On our way near the Sea Serjt Rainey & myself Saw Something verry white our curiosity was such we let our Animals Graze & went to see what it was; when we came to it there was laying before us I suppose 100 Acres of Salt about ½ an Inch deep over the Surface many places 1 ½ Inches we could Gathered barrels of it. I took about a pint for my use as beautiful as I ever saw. (Bliss 1846-7)

Private Bliss was not alone in seeing North County's salt flats in a positive light. Other 18th and 19th century observers described "white glitter" (Crespí and Brown 2001; see also Costansó and Browning 1992) and plains "as white and glistening as snow" (Rodgers 1887-8b).

For some, the salt flats' beauty may have been at least partially fueled by the significant economic potential represented by such large quantities of salt. This is reflected in the writings of Jesuit missionary Miguel del Barco, in describing a large flat mined for salt on the coast of Baja California, near Loreto:

> The salt is very white, beautiful, and pure. ...Because of its whiteness, the reflection of the sun on the salt pans is so great that it dazzles and will not allow those who go collect the salt to work. In order to undertake this maneuver, it is therefore necessary to wait until the sun is nearing sunset or else in the morning at a corresponding time. ... It is only when it has rained substantially that one cannot go fetch salt, as the salt pan fills with water and the salt softens and melts halfway. (del Barco et al. [ca. 1770]1980)

Salt was an important commodity among many Native American groups in California, and was one of the most traded items in the state (Anderson 2005, Timbrook 2007). Though Luiseño and Kumeyaay use of salt from the flats is not well documented, there are oblique mentions of their use of the salt flats, or *salinas*, for salt-gathering. A native resident interviewed by ethnographer John P. Harrington in the 1920s recalled that he "used to get salt" from San Elijo Lagoon (Harrington 1925), and Harrington also recorded words and place-names in the Luiseño lexicon relevant to collecting salt: *'éy-xllac* ("salt-gathering place"), *'eyva* ("the place betw[een] Oceanside and Encinitas where they got salt"), and *'é-'eylc* ("without salt – also sayable of a year when a salina is short of salt, for some years come when it is thus") (Harrington ca. 1930a,b). Early American settler Nathan Eaton (who came to what is now Leucadia south of Batiquitos Lagoon in 1875) also apparently gathered salt from Batiquitos Lagoon, and traded it, along with honey from his bee hives, to local Native Americans (Lamb 1977). These records, though fragmentary, are consistent with early salt extraction in other areas of California, including Baja California (Holser et al. 1981), the extensive estuarine salt flats in Ventura County, and the *salinas* (later salt ponds) of the south San Francisco Bay where the Chumash (Ventura), Ohlone (South Bay) and Euro-American residents would harvest salt (Grossinger and Askevold 2005, Beller et al. 2011).

Around the turn of the century, the salt flats were known more for their recreational rather than economic value. During the dry season, their smooth, level expanse was crisscrossed by roads (as seen on the earliest T-sheets), racetracks, and runways. In the summer, the bed of Buena Vista Lagoon was "the only straight, smooth road in the north end of the county" and residents would turn it into a racetrack for their "new-fangled Stanley Steamers, Thomas Flyers, Merry Oldsmobiles, Pope Hartfords and such" (Harmon 1967). In 1926, one resident built an airstrip on San Elijo Lagoon, giving people rides in his "newfangled" airplane on the weekends (Crimmins 1990). Boating and swimming were popular in the warm, shallow water covering the flats (Harmon 1967, Tenaglia 1999).

> The fun of racing away the summer twilights on the mud flats of Buena Vista lagoon had been a local pastime since the days of Model T Fords...
> —HOWARD-JONES 1982

By the second half of the 20th century, perspectives on North County salt flats were changing dramatically. No longer glistening, glittering, or beautiful (or lucrative), they were instead described as "aesthetically unappealing" (Welker and Patton 1995), "barren" (e.g., Crabtree et al. 1963, Mudie et al. 1976, County of San Diego 1979, Meyer 1980), and "sterile" (e.g., San Diego Regional Water Quality Control Board 1967, County of San Diego 1970).

Why the shift in aesthetic preferences? In part, perhaps, it may have been driven by the decline of the salt flats as a valuable resource, as well as a decreasing recognition of salt flat dominated estuaries as "natural" system types. The increase in pejorative language associated with salt flats also coincides with the rise of wastewater discharge into many of the lagoons and the associated water quality issues created by this practice.

> When there's seasonal rainfall, Batiquitos Lagoon in Carlsbad is a shimmering lick of water...one of the few majestic places left along Southern California's coast for shore birds and waterfowl. Often though...much of the lagoon is as parched and cracked as a vanished lake, a bleak and seemingly barren specter from the high-priced homes that overlook it.
> —TESSLER 1991

Salt flat at San Elijo Lagoon, August 1925. (OPPS Neg 91-34493, courtesy of National Anthropological Archives, Smithsonian Institution)

(1992) found that salt flats developed in areas that were inundated 15% of the time. At higher elevations that were submerged (by freshwater) 5% of the time, euryhaline conditions developed that fluctuated seasonally with freshwater input, followed by evaporation leading to salt accumulation.

Hypersalinity precluding the growth of marsh vegetation on salt flats in North County lagoons may have created a positive feedback cycle in which elevated evaporation rates in exposed, unvegetated areas magnified soil salinity, further excluding plant colonization and thus producing even more extreme hypersalinity, a process that would have favored the persistence of salt flats over time. In contrast, plant cover in vegetated areas would have shaded the soil and reduced evaporation rates, thus maintaining soil salinities at levels suitable for plant growth (Pennings and Bertness 1999). While seasonal and inter-annual variability in inlet closure status, extent of tidal inundation, and freshwater input would have resulted in large fluctuations in soil salinity, these positive feedback mechanisms likely dampened the effects of short-term variability in environmental conditions on habitat type distribution and contributed to the maintenance of relatively persistent patterns of salt flat and salt marsh.

Historical evidence in northern San Diego County supports the conclusion that patterns of salt flat/salt marsh distribution were relatively persistent on a multi-decadal scale from the late 19th through mid-20th centuries. The distribution of salt flat and salt marsh as depicted on the 1880s T-sheets broadly matches the distribution of both habitat types visible in the 1928 aerial imagery for most lagoons. Similarly, the habitat type distribution shown in the early aerial photographs is generally consistent with salt marsh/salt flat configurations evident in other early- to mid-20th century sources (e.g., oblique aerial photographs and 1930s T-sheet resurveys). Los Peñasquitos Lagoon appears to be an exception to this pattern of generally stable habitat type configurations from the late 19th through mid-20th centuries: the distribution of salt flat and salt marsh as depicted in sources spanning this period (e.g., 1880s T-sheet, 1928 aerial photos, 1950s oblique aerial photos) varies considerably. The drivers of early changes in habitat type distribution at Los Peñasquitos Lagoon are currently unknown.

Freshwater/Brackish Transitional Wetlands

Since the freshwater/brackish transitional wetlands historically found on the inland margins of each lagoon were not a primary focus of this study, we compiled relatively limited data about this habitat type compared with salt marsh and salt flat. Nevertheless, freshwater/brackish wetlands were documented by numerous historical sources (e.g., Spanish explorer accounts, USCGS T-sheets, USGS quads, landscape and aerial photographs, and USDA soil surveys) and in many cases occupied an extensive area upstream of each estuary. The wetlands were characterized by a range of vegetation types whose distribution would have varied depending on salinity, soil type, groundwater level, and other factors. Some of the plant species documented historically within this habitat type include pickleweed, saltgrass, tules, cattails, sedges, willows, and sycamores (e.g., Cooke 1849; Goldsworthy 1874b; Wilson 1883; Benton 1886; Rodgers and Nelson 1889; Chase 1913; Storie and Carpenter 1929a,b; Purer 1942; Crespí and Bolton 2001).

Ecological Functions

Northern San Diego County lagoons historically supported a variety of native wildlife, including a number of fish species and resident and migratory birds. Early observers described abundant waterfowl in the lagoons around the turn of the 20th century: "splendid duck shooting on the sloughs" of San Dieguito Lagoon (Sherman 2001), "countless little lagunas alongshore, often filled with ducks" (Holder 1906), and so many ducks in Batiquitos Lagoon that "water space was limited" (O'Connell 1987). Resident Richard Lyman recalled eating "a lot of plump, roasted pinwheel ducks" from Batiquitos Lagoon in the early 1900s (Hasket 1999). Salt marshes supported feeding, breeding, and refuge for resident birds, such as the state-endangered Belding's savannah sparrow *(Passerculus sandwichensis beldingi)* and the state- and federally endangered light-footed Ridgway's rail *(Rallus obsoletus levipes;* formerly light-footed clapper rail, *Rallus longirostris levipes).* Freshwater and brackish wetlands at the inland edges of lagoons likely supported many additional species such as California red-legged frog *(Rana draytonii),* two-striped garter snake *(Thamnophis hammondii),* bears, coyotes, and deer.

The functions provided to birds and other wildlife shifted throughout the year, varying with connectivity to the ocean and the depth of inundation of the salt flats. During the late spring and summer, the drying salt flats offered breeding habitat for birds such as the state- and federally endangered California least tern *(Sterna antillarum browni)* and federally threatened western snowy plover *(Charadrius nivosus nivosus)* (see page 160). Birds such as the Belding's savannah sparrow would have used the salt flats to forage, and migratory birds would have used them for resting. Salt flats also provided habitat for invertebrates such as tiger beetles *(Cicindela* spp.) and rove beetles *(Bledius* spp.) and corridors for traveling mammals (Welker and Patton 1995, Desmond et al. 2001).

During periods of shallow inundation – such as in the rainy season as closed lagoons began to fill with freshwater flow and precipitation, or during periods of tidal exchange when portions of the mud flats and salt flats were shallowly flooded by the tides – the salt flats would have provided habitat for a wide range of migratory waterbirds (Taft et al. 2002). Areas with four to six inches of water, and gradients of shallower water near the lagoon edges, would have provided ideal foraging habitat for American avocet *(Recurvirostra americana),* black-necked stilt *(Himantopus mexicanus),* white-fronted goose *(Anser albifrons),* snow goose *(Chen caerulescens),* a range of herons and egrets (Ardeidae), multiple species of plovers (Charadriianae), sandpipers (Scolopacidae), and many other shorebirds (Baker 1979, Taft et al. 2002). Water between two and ten inches depth would have been used extensively by dabbling ducks, including many migrating and wintering species such as gadwall *(Anas strepera),* American wigeon *(A. americana),* northern pintail *(A. acuta),* blue-winged teal *(A. discors),* and cinnamon teal *(A. cyanoptera;* Isola et al. 2000, Taft et al. 2002). In lagoons with relatively deep water (>10 inches), flooded areas would have been used as foraging habitat by diving birds such as grebes (Podicipedidae), cormorants *(Phalacrocorax* spp.), ruddy duck *(Oxyura jamaicensis),* canvasback *(Aythya valisineria),* redhead *(A. americana),* bufflehead *(Bucephala albeola),* and several species of mergansers (Merginae), while dabbling ducks and shorebirds would have used shallower areas along the edges (Isola et al. 2000, Taft et al. 2002). The interface between mud flat, marsh, salt flat, and open water would have provided diverse areas for cover and feeding (Welker and Patton 1995).

These systems also would have supported a number of euryhaline fish species (adapted to be able to withstand broad range of salinities), contributing to the region's biodiversity (Williams and Zedler 1999). The San Dieguito River historically supported federally endangered southern California steelhead (*Oncorhynchus mykiss*; NMFS 2012); these fish likely used the estuary as rearing habitat as has been shown for other California estuaries (Bond 2006, Hayes et al. 2008). Of particular interest, intermittently closing estuaries provide habitat for the federally endangered tidewater goby (*Eucyclogobius newberryi*), a fish only found in these types of estuarine systems (see page 161).

Today, the lagoons continue to support a range of estuarine and other wetland species, including the threatened and endangered Belding's savannah sparrow, light-footed Ridgway's rail, California least tern, western snowy plover, and riparian species including least Bell's vireo (*Vireo bellii pusillus*) and southwestern willow flycatcher (*Empidonax traillii extimus*; Caltrans and SANDAG 2013). However, changes to the lagoons, particularly beginning in the mid-20th century, have heavily reduced the value of many of these systems for wildlife. Habitat loss and type conversion have occurred through dredging and filling, alterations to freshwater flows, changes in tidal hydrology, and increased sedimentation from development. Other activities have impacted the quality of habitat, including water quality degradation (e.g., from effluent discharge) and disruptive human uses. For example, in

The coast of Southern California is, in the main, a long stretch of sand dunes changing every hour and moment in the wind that heaps them up into strange and fascinating shapes. In many instances they form breakwaters, damming up the waters that flow down the cañons' streambeds from the interior…At Alamitos, where the San Gabriel River reaches the sea, and at Balsa Chica, one of the finest preserves and clubs in the country, and other places along shore to San Diego we shall find these lagunas, or sea swamps, the home of the duck, goose, and swan.

— HOLDER 1906

Ducks of nearly all varieties were found in every lagoon and slough. …
The sloughs and bays along the coast were lined with curlew, snipe, willet, dowitchers, plover, etc.

—VAN DYKE ET AL. 1888, REFERRING TO THE SAN DIEGO COUNTY COASTLINE

the 1970s the eastern end of Batiquitos Lagoon was used by recreational vehicles and for helicopter landing, "with the result that the salt flat is criss-crossed by vehicle tracks and noise is often excessive" (County of San Diego 1979). In general, land use impacts and management activities have tended to reduce seasonal and interannual variability, shift or compress physical gradients, and decrease habitat complexity and heterogeneity, affecting the ecological functions provided by the lagoons and in many cases shifting the suites of supported species (e.g., from brackish to marine fish communities).

A flock of approximately 300 Wood Ibises was noted during July and August, 1953, just south of Oceanside in the Buena Vista Lagoon…[this] is heartening to ornithologists who have watched with much anxiety the encroachment of commercial, recreational and flood control development in the slough, lagoon and shallow bay areas of southern California during recent years. As the available feeding grounds face severe reduction due to such development, we may be on the eve of seeing fewer, instead of more, of these American storks.

—RECHNITZER 1954

Waterbirds in San Elijo Lagoon, January 2013. (photo by Sean Baumgarten)

ECOLOGICAL FUNCTIONS BY HABITAT TYPE

Northern San Diego County lagoons historically supported a variety of native wildlife (see page 155). This spread illustrates examples of key functions likely provided by each habitat type. Scientific names are provided for species not discussed elsewhere in this chapter.

CHANNELS AND PONDS

Feeding, breeding, and refuge habitat for resident and migratory birds

Feeding, breeding, and refuge habitat for resident fish (e.g., Pacific staghorn sculpin (*Leptocottus armatus*), California killifish (*Fundulus parvipinnis*), and federally endangered tidewater goby)

Feeding, spawning, nursery, and refuge habitat for marine fish (e.g., topsmelt (*Atherinops affinis*)), depending on inlet closure status

Intertidal mud flat habitat for invertebrates (e.g., crabs, mollusks, shrimp)

light-footed Ridgway's rail

EMERGENT SALT MARSH

Feeding, breeding, and refuge habitat for resident birds (e.g., Belding's savannah sparrow, light-footed Ridgway's rail) and transient and resident mammals and reptiles

Primary production supports invertebrates, resident and marine fish, and resident and migratory birds

tidewater goby

Pacific staghorn sculpin

Belding's savannah sparrow

Tidewater goby: Josh Hull, courtesy U.S. Fish and Wildlife Service; Pacific staghorn sculpin: Jonathan Klenk, courtesy Calfish, UC Davis; Belding's savannah sparrow: Matt Sadowski, courtesy U.S. Fish and Wildlife Service; License: https://creativecommons.org/licenses/by/2.0/.

ocean   dunes   channels and ponds   salt marsh

SEASONALLY FLOODED SALT FLAT

Inlet closed, lagoon drying out

Foraging, breeding, and resting habitat for migratory birds (e.g., California least tern, western snowy plover) and resident birds (e.g., American avocet, black-necked stilt)

Habitat for invertebrates such as tiger beetles and rove beetles

Inlet closed, lagoon filling

Feeding and resting habitat for migratory and resident waterbirds (diving ducks, terns, and cormorants in deeper areas; dabbling ducks and wading birds at margins)

Inlet open

Feeding and resting habitat for migratory and resident waterbirds (dabbling ducks, shorebirds, terns, and cormorants)

marsh wren

FRESHWATER/BRACKISH WETLANDS

Feeding, breeding, and sheltering habitat for wetland and riparian species (e.g., California red-legged frog, two-striped gartersnake, marsh wren (*Cistothorus palustris*), least Bell's vireo, southwestern willow flycatcher)

Refuge for birds when marsh and salt flat flooded

Primary production supports invertebrates and resident and migratory birds

California least tern (juvenile)

California red-legged frog

two-striped gartersnake

common goldeneye

California least tern (juvenile): courtesy Linda Tanner; Common goldeneye: Maga-Chan, courtesy Wikipedia Commons; California red-legged frog: Flo Gardipee, courtesy U.S. Fish and Wildlife Service; two-striped garter snake: courtesy U.S. Geological Survey; marsh wren: Cephas, courtesy Wikipedia Commons; License: https://creativecommons.org/licenses/by/2.0/.

seasonally flooded salt flat   freshwater/brackish wetland   upland

NESTING ON THE SALT FLAT:
CALIFORNIA LEAST TERNS AND WESTERN SNOWY PLOVERS

Federally endangered California least terns *(Sterna antillarum browni)* and federally threatened western snowy plovers *(Charadrius nivosus nivosus)* both used the seasonally dry salt flats of North County lagoons as breeding habitat. On the west coast, snowy plovers and least terns breed in the late spring through summer (Jacobs 1986, Akçakaya et al. 2003), and have been documented to use unvegetated salt flats as nesting sites (Jacobs 1986, Koenen et al. 1996).

The presence of least terns and snowy plovers nesting at northern San Diego County beaches and estuaries was well documented in the early part of the 20th century. Between 1921 and 1945, specimen collectors recorded over 20 instances of least tern eggs and over 20 instances of snowy plover eggs and nests. While many of the collection sites can be difficult to precisely locate, they included Batiquitos and San Dieguito lagoons (e.g., Harrison 1932a,b, 1934a,b, 1945; G. Bancroft Collection 1932; Carpenter 1939) and possibly San Elijo Lagoon (Heaton 1923). In the later part of the 20th century, the extensive salt flats still present at a number of lagoons were recognized as habitat for the least tern and snowy plover, which were observed nesting on the flats at San Elijo and Batiquitos lagoons (County of San Diego 1979, Welker and Patton 1995).

California least tern chick (above). (R. Baak, courtesy of U.S. Fish and Wildlife Service)
California least tern (below). (Mark Pavelka, courtesy of U.S. Fish and Wildlife Service)

Tidewater goby. (courtesy of U.S. Fish and Wildlife Service)

TIDEWATER GOBY

The tidewater goby *(Eucyclogobius newberryi)* is a federally endangered fish uniquely adapted to life in the intermittently closing estuaries of California's coast. Tidewater gobies are found exclusively in California's coastal brackish-water habitats, rather than in fully tidal or freshwater systems (Swift et al. 1989, Capelli 1997). Tidewater gobies prefer shallow water with a relatively low salinity (under 20 ppt; Swift et al. 1989, Capelli 1997), though they can tolerate large fluctuations in salinity (USFWS 2007); these conditions are often found in systems that are seasonally separated from the ocean and do not experience continuous tidal flushing (Capelli 1997).

Because of their relatively limited marine dispersal ability (Lafferty et al. 1999), tidewater gobies exhibit substantial local and regional differentiation as shown by genetic (Dawson et al. 2001) and morphological (Ahnelt et al. 2004) studies. The southern tidewater goby populations found in San Diego County are dramatically distinct from populations found further to the north, with differences that appear to justify classification as a separate species (Ahnelt et al. 2004, Earl et al. 2010). Since the southern tidewater goby are currently only found in northern San Diego County (at Camp Pendleton; Dave Jacobs, pers. comm.), this distinction carries significant implications for the management of potential tidewater goby habitat in North County lagoons.

Tidewater gobies were last documented in the six lagoons studied here in 1940 (Agua Hedionda Lagoon; Miller and Miller 1940) and 1953 (Buena Vista Lagoon; USFWS 2005; Camm Swift, pers. comm.). Agua Hedionda Lagoon, which was designated as critical habitat for the tidewater goby in 2000 (USFWS 2005), also represents the southernmost recorded observation of a tidewater goby. The absence of tidewater goby observations from intermittently closing estuaries south of Agua Hedionda Lagoon may indicate their historical absence from these systems, or may reflect their extirpation prior to the mid-20th century as a result of anthropogenic modification (Capelli 1997, USFWS 2005). The latter hypothesis is tentatively supported by an account discovered during this study from Rechnitzer (1956), who noted that tidewater goby were present in the marsh channels of San Elijo Lagoon along with arrow goby (*Clevelandia ios*), shadow goby (*Quietula y-cauda*), and other fish, and that "carcasses of all those fishes, except the shadow goby, were found along the banks of the watercourses following a feeding foray by the ibis [wood stork, *Mycteria americana*]." However, Rechnitzer's observation is unconfirmed by specific museum records.

Habitat Type Change Analysis

Comparing historical and contemporary habitat patterns yields a number of insights useful to managers and restoration planners. At the regional level, our findings reveal substantial shifts in habitat type extent and distribution, and highlight how land use changes have resulted in ecological change. At an individual lagoon level, the results show the relative changes in the extent of each habitat type and identify particular areas where remnant features have persisted or novel habitat types have emerged. These insights can be useful when considering the palette of habitat types to be included in restoration efforts.

This section summarizes shifts in habitat type distribution for all six lagoons within the study area. More detailed change analysis for individual lagoons is presented within each lagoon chapter (see chapters 4-9).

Methods

We used contemporary wetland and vegetation mapping to evaluate how habitat type extent and distribution across the six lagoons has changed over the past ~200 years. Contemporary wetland mapping layers for the southern California region based on the Cowardin classification system (CSUN Center for Geographical Studies 2012) were used as the base layer for contemporary conditions. (These layers were developed based on aerial imagery flown prior to the most recent lagoon modifications, such as the San Dieguito Wetlands Restoration Project). Supplementary local vegetation mapping was used where available in cases where further distinction between the habitat types shown in the regional mapping was necessary (as described in more detail below). Additional local mapping was obtained for Buena Vista Lagoon (Everest International Consultants, Inc. 2004), San Elijo Lagoon (AECOM 2012), and Los Peñasquitos Lagoon (Greer and Stow 2003). Modern aerial imagery (NAIP 2009) was also used to verify classifications in the vegetation mapping. The combined contemporary mapping is referred to using an approximate date of ca. 2010.

The wetland extent used in this analysis included the historical estuarine extent of each lagoon in addition to any areas outside of the historical lagoon footprint that were mapped as estuarine wetlands in the contemporary regional wetland mapping. We also included the area of freshwater/brackish transitional wetland habitats mapped upslope of each lagoon within the analysis footprint in order to capture changes in these features. Note, however, that this analysis does not always capture the full extent of upslope freshwater/brackish wetlands in the contemporary landscape.

We developed a crosswalk to enable comparison between the different habitat classification systems used in the historical and contemporary mapping (table 10.2 at right). This process involved grouping some contemporary classes to make the two mapping products comparable (e.g., subtidal open water and intertidal mud flat were grouped together to match the resolution used in the historical mapping). We also made two additional distinctions in the contemporary regional wetland mapping classes to facilitate comparison with the habitat types used in the historical mapping. First, we separated salt flats and mud flats (which are grouped together in the Cowardin classification system used in the contemporary mapping) based on the presence or absence of a connection to subtidal open water. Second, we distinguished brackish marsh and salt marsh (also grouped together in the Cowardin classification system), using additional local vegetation mapping and modern aerial imagery for individual systems. This was necessary because in the historical mapping brackish marsh is grouped with freshwater marsh, rather than with salt marsh.

Table 10.2. Crosswalk between historical and contemporary habitat classifications. Classifications used in contemporary regional wetland mapping were crosswalked to historical habitat classifications. Contemporary regional wetland mapping (CSUN Center for Geographical Studies 2012) is based on the Cowardin classification system (Cowardin et al. 1979). In most cases, the greater level of resolution in the contemporary mapping meant that multiple contemporary classes were crosswalked to a single historical classification. In several cases, however, a single contemporary class was split into two classes to allow for comparability to historical mapping; additional local vegetation mapping (Greer and Stow 2003; Everest International Consultants, Inc. 2004; AECOM 2012) was used to provide further distinction between contemporary habitat types where necessary. A number of features without a direct historical/modern comparison (e.g., artificial rocky shore), encompassing a total of only 58 acres, were placed into the category "Other." A separate crosswalk (not shown here) was used to match the historical habitat classifications with contemporary vegetation community types included in AECOM (2012) mapping for San Elijo Lagoon.

HISTORICAL CLASSIFICATION	CONTEMPORARY CLASSIFICATION		
	System/Subsystem	Class	Wetland Code
Open water/mud flat	Estuarine subtidal (E1)	Unconsolidated Bottom (UB)	E1UB, E1UBh, E1UBLh, E1UBx
Open water/mud flat	Estuarine intertidal (E2)	Streambed (SB)	E2SB, E2SBh
Open water/mud flat	Estuarine intertidal (E2)	Aquatic Bed (AB)	E2ABh
Open water/mud flat[1]	Estuarine intertidal (E2)	Unconsolidated Shore (US)	E2US, E2US/EMh, E2USh, E2USx
Salt flat (seasonally flooded)[2]	Estuarine intertidal (E2)	Unconsolidated Shore (US)	E2US, E2US/EMh, E2USh, E2USx
Salt marsh[3]	Estuarine intertidal (E2)	Emergent (EM)	E2EM, E2EMh, E2EMx
Freshwater/brackish wetland	Palustrine	Unconsolidated Bottom (UB)	PUB, PUBh
Freshwater/brackish wetland[4]	Estuarine intertidal (E2)	Emergent (EM)	E2EM, E2EMh, E2EMx
Freshwater/brackish wetland	Estuarine intertidal (E2)	Scrub-Shrub (SS)	E2SS, E2SSh
Freshwater/brackish wetland	Estuarine intertidal (E2)	Scrub-Shrub/Emergent (SS/EM)	E2SS/EM, E2SS/EMh
Freshwater/brackish wetland	Palustrine	Emergent (EM)	PEM, PEMh, PEMx
Freshwater/brackish wetland	Palustrine	Forested (FO)	PFO, PFO/EMh, PFO/EMx, PFO/SS, PFOx
Freshwater/brackish wetland	Palustrine	Scrub-Shrub (SS)	PSS, PSSh
Freshwater/brackish wetland	Palustrine	Scrub-Shrub/Emergent (PSS/EM)	PSS/EM, PSS/EMh, PSS/EMx, PSS\EMh
Freshwater/brackish wetland	Palustrine	Unconsolidated Shore (US)	PUS, PUSx
Freshwater/brackish wetland	Riverine tidal (R1)	Emergent (EM)	R1EMx
Developed	UNCLASSIFIED		
Other	Estuarine intertidal (E2)	Rocky Shore (RS)	E2RSr
Other	Palustrine	Unconsolidated Bottom (UB)	PUBrx, PUBx
Other	Riverine tidal (R1)	Unconsolidated Bottom (UB)	R1UB, R1UBx
Other	Riverine lower perennial (R2)	Aquatic Bed (AB)	R2ABx
Other	Riverine lower perennial (R2)	Unconsolidated Bottom (UB)	R2UB, R2UBrx, R2UBx
Other	Riverine lower perennial (R2)	Unconsolidated Shore (US)	R2US
Other	Riverine upper perennial (R3)	Unconsolidated Bottom (UB)	R3UB
Other	Riverine intermittent (R4)	Streambed (SB)	R4SB, R4SBrx, R4SBx
Other	Marine intertidal (M2)	Unconsolidated Shore (US)	M2US

[1]Distinguished from salt flat (seasonally flooded) based on presence of a connection to subtidal open water.

[2]Distinguished from open water/mud flat based on absence of a connection to subtidal open water.

[3]Distinguished from freshwater/brackish wetland based on additional local vegetation mapping and modern aerial imagery.

[4]Distinguished from salt marsh based on additional local vegetation mapping and modern aerial imagery.

One of the challenges in analyzing changes in habitat type distribution is that complex, heterogeneous habitat types must necessarily be combined into a limited number of broad classifications to reflect the resolution of available data and fit both historical and contemporary mapping into comparable categories. As a result, not all shifts in habitat type distribution are captured in this analysis. For instance, a broad range of vegetation communities, characterized by species as diverse as saltgrass, cattails, and riparian trees, are included within the freshwater/brackish wetland classification, and thus even very significant conversions between these types (e.g., from a saltgrass-dominant community to a cattail-dominant community) is not captured in the change analysis. Similarly, open water was historically confined to shallow channels and ponds, while today most of the open water exists as large subtidal basins. Though shallow channels and large subtidal basins represent very different habitats and support different suites of species, both environments are grouped together under the open water/mud flat habitat type.

Results

The modern habitat mosaics that characterize northern San Diego County's lagoons are in many cases quite different from the array of habitats that dominated the lagoons historically. Multiple changes in land and water use have impacted the lagoons both directly and indirectly over the past 150 years (see Chapter 3), substantially altering habitat type distributions and with significant implications for the ecological functions provided by the lagoons.

In some cases, such as San Elijo Lagoon, much of the overall historical habitat configuration is intact. In other cases, the modern lagoons bear little resemblance to their historical counterparts. For example, Buena Vista Lagoon, historically dominated by an extensive central salt flat, is now characterized by open water basins surrounded by emergent freshwater/brackish vegetation, while the salt flats of Batiquitos and Agua Hedionda lagoons have been replaced by deeper subtidal open water features. Even in highly altered systems, however, remnant features such as channels and ponds can still be found.

Although each lagoon has experienced a distinct trajectory (see chapters 4-9), it is instructive to review overall trends across the six systems (table 10.3 below and graph at right). There has been an overall loss of total estuarine area resulting from the substantial filling and development that has occurred within some of the lagoons. The vast majority of the seasonally flooded salt flat habitat type, which historically covered extensive areas in five

Table 10.3. Absolute and percentage change in habitat type extent across all six lagoons. Acreages are rounded to the nearest 10 acres, and percentages are rounded to the nearest 1% (acreages and percentages may not agree exactly due to rounding).

	Historical (acres)	Contemporary (acres)	% Change
Salt marsh	1,330	1,170	-12%
Salt flat (seasonally flooded)	1,230	120	-90%
Open water/mud flat	140	980	615%
Freshwater/brackish wetland	1,650	760	-54%
Developed		1,440	

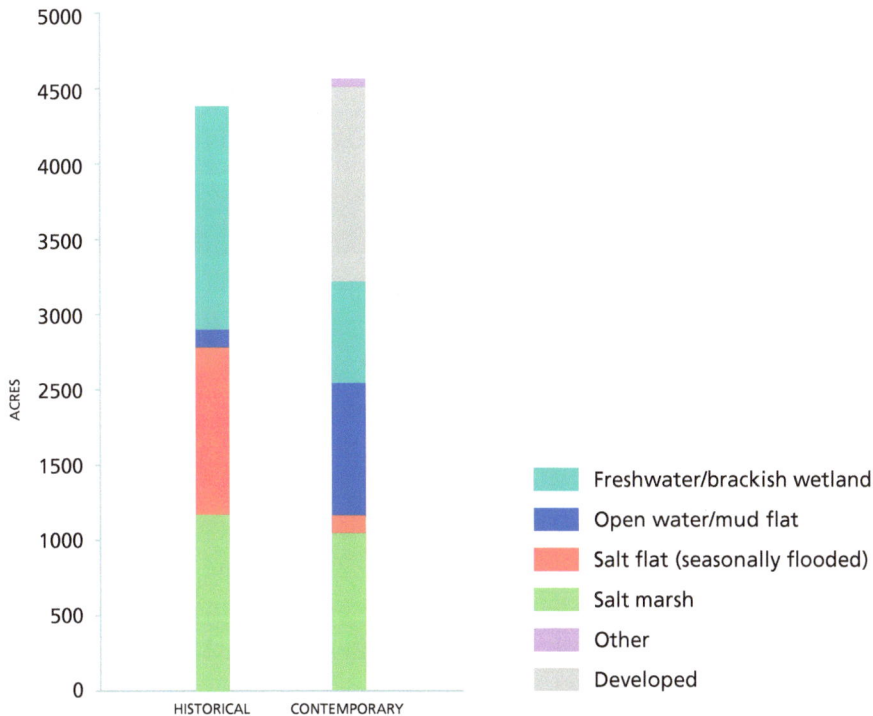

Change in habitat type distribution across all six lagoons. The geographic extent of this analysis includes the historical footprint of the lagoon and adjacent freshwater/brackish transitional wetland areas, as well as additional features classified as estuarine in the contemporary mapping. Comparison of past and present conditions reveals the loss of salt flat, salt marsh, and freshwater/brackish wetlands with a concomitant gain in open water/mud flat. The slight difference in the height of the two bars is due to the relatively small extent of additional estuarine features included in the contemporary mapping that were not wetland features historically. (data from Greer and Stow 2003; Everest International Consultants, Inc. 2004; NAIP 2009; AECOM 2012; CSUN Center for Geographical Studies 2012)

of the lagoons, has largely been converted to other habitat types. Salt marsh acreage has decreased by a relatively small amount, though much of the contemporary salt marsh exists in areas that did not support salt marsh historically. Conversely the open water/mud flat habitat type, which was historically limited to small channel networks and ponds within the salt marshes, has increased in extent across all six lagoons and has become a dominant habitat type in the northern three lagoons. Most of the freshwater/brackish wetlands that historically occupied the valleys upstream of the lagoons have been lost to development; however, freshwater/brackish wetlands have also expanded downstream into areas historically occupied by other estuarine habitat types. The following paragraphs describe these trends in more detail.

Historically covering more than 1,200 acres, today salt flat ccupies just 120 acres (~10% of the historical extent). Conversion to open water and intertidal mud flat accounts for the largest portion of the lost salt flat acreage (~42%); large amounts of salt flat have also been replaced by salt marsh, freshwater/brackish wetland, and developed areas. The conversion of salt flat to open water/mud flat was in large part the result of direct manipulation. For example, dredging at Agua Hedionda (1950s) and Batiquitos (1990s) lagoons eliminated nearly all of the historical salt flat area and increased the area of subtidal open water and intertidal mud flat by approximately 260 acres in each of the lagoons (Ritter 1963; Merkel & Associates, Inc. 2009).

In contrast, the open water/mud flat category now occupies more than seven times the area that it did historically as a result of dredging and other manipulations. This increase has been especially significant in the northern three lagoons, where open water and mud flat have increased by roughly 1,000-3,000% (compared with an approximately 150-200% increase in the southern three lagoons). The expansion of open water habitat has been driven largely by the creation of large subtidal basins at Agua Hedionda, Batiquitos lagoons, and San Dieguito lagoons, and by the construction of the weir at Buena Vista Lagoon. As a result, whereas the areas of perennial open water in the historical lagoons were generally confined to narrow channels and small, shallow ponds, a substantial portion of the contemporary open water area is comprised of deeper subtidal habitat.

Salt marsh was one of the dominant habitat types historically within the lagoon complexes, though the historical extent of salt marsh varied considerably by lagoon, ranging from less than 75 acres at Buena Vista Lagoon to over 540 acres at San Dieguito Lagoon. The acreage of salt marsh has decreased slightly across the six lagoons, falling from approximately 1,330 acres to just under 1,170 acres (~12% decrease). Loss of historical salt marsh has been driven by development and conversion to other habitat types, such as open water. Buena Vista, Agua Hedionda, San Dieguito, and Los Peñasquitos lagoons have all experienced a sizeable loss of salt marsh acreage (~60-180 acres each), while Batiquitos and San Elijo Lagoons have actually seen a considerable increase in salt marsh (~60-190 acres each).

While the total decrease in salt marsh extent is relatively small, there has been a substantial shift in the location of this habitat type within many of the lagoons. Overall, only about 30% of the salt marsh present today falls within the historical salt marsh footprint. The remaining 70% is found in areas where salt marsh was not documented historically, especially in the central and eastern portions of lagoons in areas formerly occupied by salt flats or freshwater/brackish wetlands. Some of this shift appears to reflect wetland creation activities, though it is possible that some of the high salt marsh areas represented in the contemporary mapping are similar to the more brackish components of the freshwater/brackish wetland complexes shown in the historical mapping. San Elijo, San Dieguito, and Los Peñasquitos lagoons have each lost ~25-50% of their historical salt marsh, while Buena Vista, Agua Hedionda, and Batiquitos lagoons have each lost ~80-100% of their historical salt marsh area.

Another important change in habitat type distribution has been the significant decrease in the extent of freshwater/brackish wetlands. These wetlands historically occupied over 1,600 acres within the valleys adjacent to the lagoon complexes, but today they occupy less than 800 acres (these figures account for only those freshwater/brackish wetlands within the historical mapping footprint; additional contemporary freshwater/brackish wetlands exist outside of this historical footprint but were not included in the analysis). Urban development (such as the construction of homes, shopping areas, roads, and golf courses) and other direct modifications eliminated over 50% of these wetlands. The exceptions to this trend are Los Peñasquitos and San Elijo lagoons, where the extent of freshwater/brackish wetlands has actually increased due to significant expansion of this habitat type into

Interface between freshwater marsh (at right) and salt marsh (at left) in Los Peñasquitos Lagoon, December 2012. (photo by Robin Grossinger)

areas historically dominated by salt marsh or seasonally flooded salt flat. The expansion of freshwater/brackish wetlands in these areas appears to have been driven by a combination of factors. Wastewater discharge, agricultural irrigation, and urban runoff have all contributed to increased freshwater inputs over the past 50-100 years, likely reducing salinity levels on the inland margins of these lagoons. In addition, urban development within the watersheds has increased sediment delivery to the lagoons, while inlet constriction due to road and railroad construction has likely increased sediment retention, potentially raising elevations in portions of the lagoons above the range that will support salt marsh or salt flats (Welker and Patton 1995, Cole and Wahl 2000, Greer and Stow 2003, San Elijo Lagoon Conservancy 2005, White and Greer 2006).

Urban development has encroached on other habitat types in addition to freshwater and brackish wetlands: approximately 140 acres of seasonally flooded salt flat and 410 acres of salt marsh have been replaced by developed areas (about 11% and 31% of the historical extent of these habitat types, respectively). In total, roughly 33% percent of the historical footprint of the lagoons and the adjacent freshwater/brackish wetlands are now developed. This trend varies widely by lagoon, however. For instance, approximately 50% of San Dieguito Lagoon's historical footprint is now developed, while only about 10% of Los Peñasquitos Lagoon's historical footprint is developed.

Past and Present Comparison

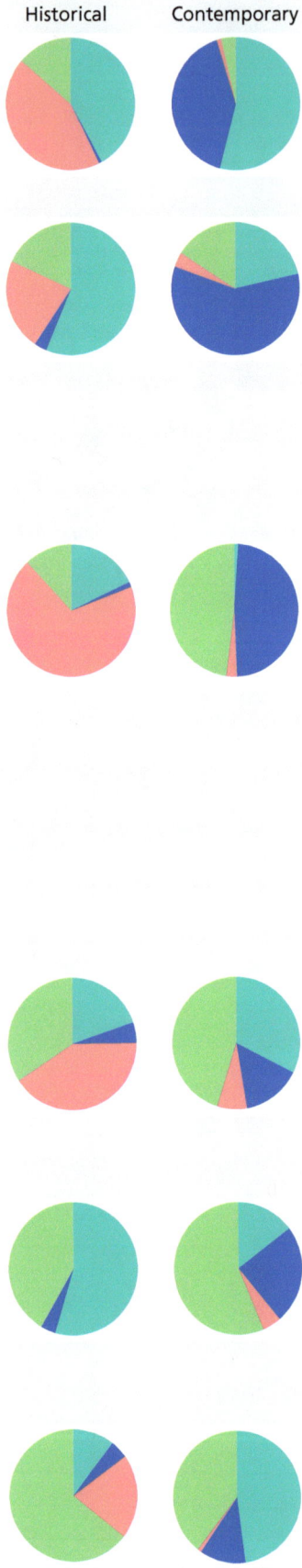

Historical   Contemporary

Contemporary Habitats: Change from Past Conditions

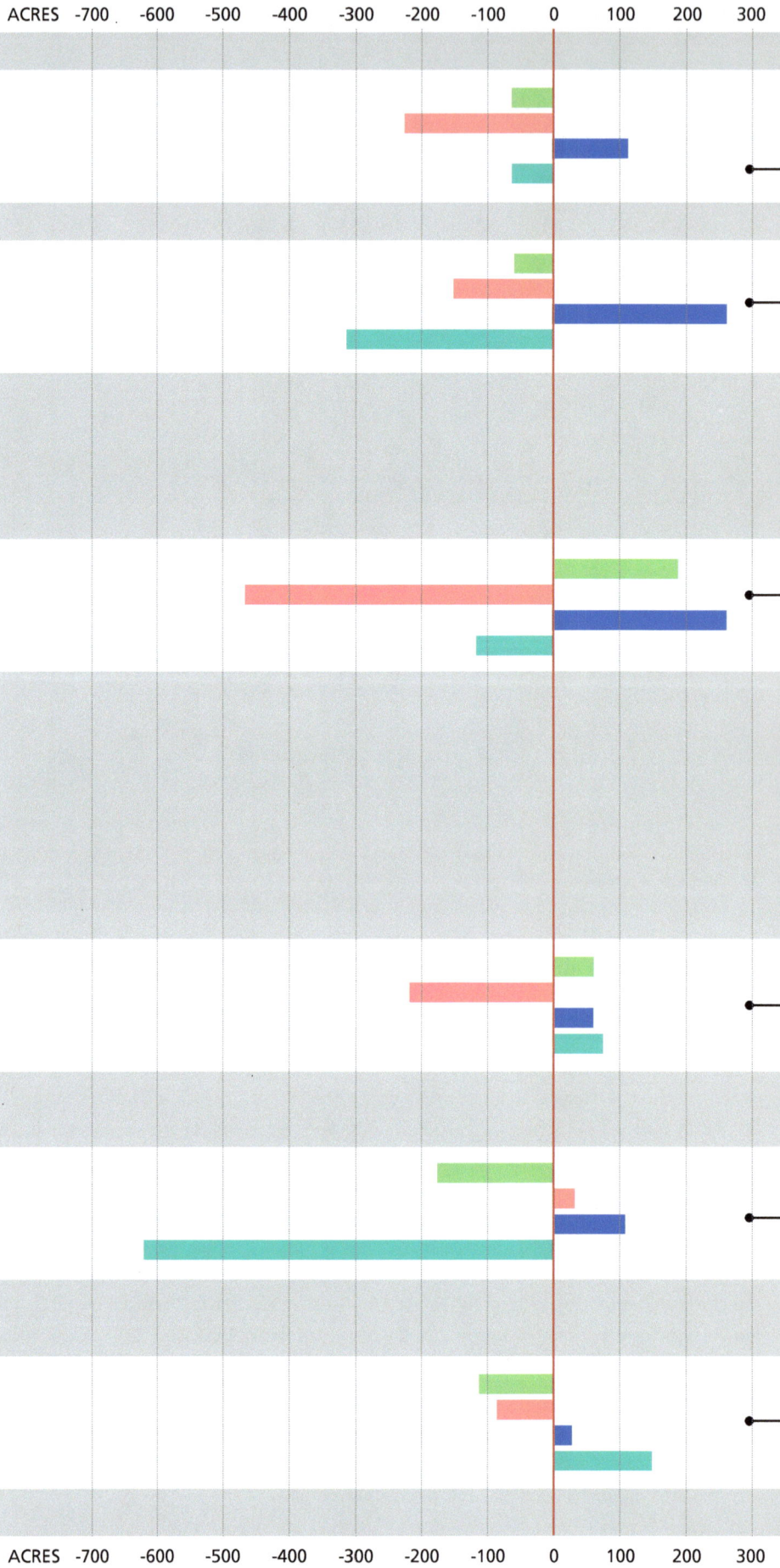

ACRES -700 -600 -500 -400 -300 -200 -100 0 100 200 300

Salt Marsh
Salt Flat (Seasonally Flooded)
Open Water / Mud Flat
Freshwater / Brackish Wetland

ACRES -700 -600 -500 -400 -300 -200 -100 0 100 200 300

Oceanside

Vista

Buena Vista Lagoon

Carlsbad

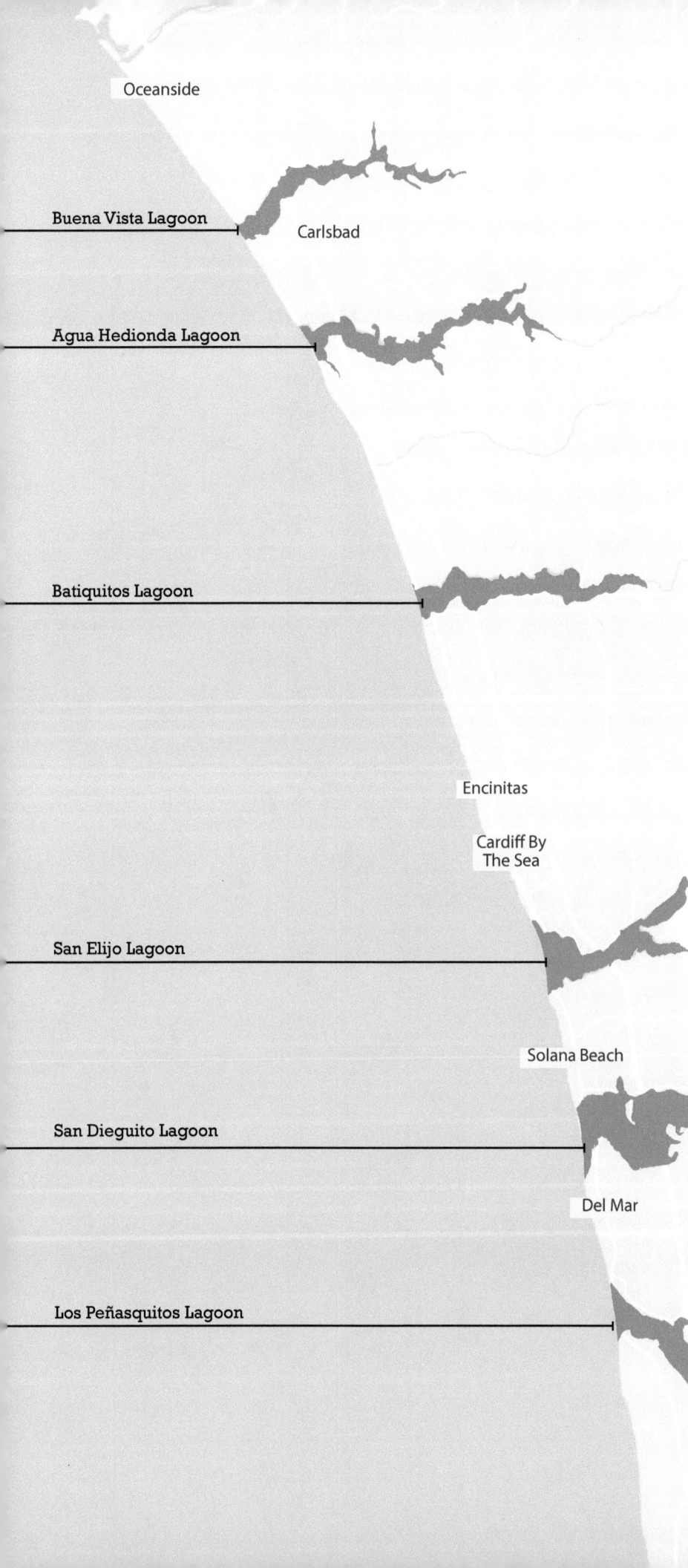

CHANGE OVER TIME

Northern San Diego County's coastal lagoons have been subject to many changes over the past 150 years. The lagoons were historically dominated by a combination of salt marsh and seasonally flooded salt flats, with smaller amounts of open water and mud flat confined to shallow ponds and channels. Often extensive freshwater/brackish wetlands bordered the lagoons on their inland margins. Though North County lagoons have only lost about 15% of their total estuarine area (not including additional losses in freshwater/brackish wetlands), today many of the lagoons support a much different array of habitats as a result of urbanization, filling, dredging, wastewater discharges, and other land use changes. Most of the historical salt flat area has been eliminated, replaced by marsh or subtidal basins. In turn, large open water basins have been created in several of the lagoons in areas historically occupied by salt marsh or salt flat. Extensive areas historically covered by freshwater/brackish wetlands have been lost to development, though in some cases these wetlands have also expanded into areas that historically supported estuarine habitat types.

Agua Hedionda Lagoon

Batiquitos Lagoon

Encinitas

Cardiff By
The Sea

San Elijo Lagoon

Solana Beach

San Dieguito Lagoon

Del Mar

Los Peñasquitos Lagoon

*R*econstructing the historical ecology of North County's estuaries requires an understanding not just of the ecological patterns that characterized the lagoons historically but also an understanding of the dynamic physical processes that operated across these systems. We first examine the two key watershed processes — freshwater flows and sediment inputs — that influenced lagoon morphology and ecology, and discuss the impacts of historical sediment accretion and sea level rise rates on lagoon evolution. We then analyze several lines of evidence to characterize historical inlet closure dynamics, including historical observations, tidal prism calculations, and core data. Finally, we present a conceptual synthesis that illustrates the connections between seasonally variable physical processes and historical ecological patterns.

Watershed Dynamics

Though an exhaustive analysis of historical fluvial dynamics across northern San Diego County's watersheds is outside the scope of this study, a brief overview of freshwater and sediment inputs based on early maps and textual descriptions is presented here. (See chapters 4-9 for additional local detail on watershed hydrology and page 36 for information about changes in watershed hydrology over time.)

Freshwater Inputs

Historical streamflow is difficult to quantify, as nearly all streams in the study area were not gaged before the major modifications of the 20th century. In addition, surface flow patterns varied by stream, reach (e.g., canyons versus alluvial plains; upper watershed versus lower watershed), and year, making generalizations difficult.

In general, streamflow entering northern San Diego County lagoons was highly seasonally variable. The majority of freshwater flow and sediment reaching the lagoons was conveyed during the wet season, when high flows created direct connections between creeks and lagoons. During periods of prolonged or heavy rainfall, creeks would swell: for example, when travelling from San Diego north across the study area Spanish explorer de Anza found that the "creeks have become rivers" after steady rainfall in February 1776, and he had difficulty crossing them (de Anza and Bolton 1930). Similarly,

(top) **San Dieguito River and Del Mar Fairgrounds, ca. 1954.** (Collection 87-26, courtesy of Scripps Institution of Oceanography Archives, UC San Diego)

travelers in winter 1869 noted that creeks at Los Peñasquitos, San Dieguito, and San Elijo were "full and strong" and hard to cross (Bell 1869). Even in winter flow could be limited, however, as a newspaper article describing a journey from San Luis Rey to San Diego in January 1884 attests (this trip was taken just before the extreme flooding of February 1884, which caused extensive damage to coastal roads and the railway):

> As we crossed these valleys near the ocean we found the creek beds without water, which is, I believe, not uncommon throughout the county, as the water prefers to hide itself in the sand before reaching the ocean in every stream except San Luis Rey river. (*Pacific Rural Press* 1884)

In the summer, small watershed size coupled with low rainfall resulted in relatively limited freshwater inputs to most of the lagoons. Archival data suggest that most of the region's streams maintained extensive intermittent reaches that retained little or no surface water during the dry season, particularly as they ran through broad alluvial valleys. This is reflected in general descriptions of regional patterns in dry-season surface flow – for example, Hall (1888) wrote that the region's streams "seldom flow continuously to the sea during the dry months of summer and fall, except following seasons of unusual rainfall." Records from the downstream ends of creeks as they approached the lagoons support this generalization, for example documenting water in pools but no surface flow (Agua Hedionda Creek in

WATER QUALITY ATTRIBUTES

Historically, water quality attributes in the study systems would have varied widely depending on eutrophication, oxygen dynamics, and hypersalinity associated with wetting and drying patterns. During periods when the estuary mouths were open and the systems experienced regular tidal flushing, low oxygen and high salinity conditions would have been rare. In contrast, during periods when closed mouths restricted access to the tides, the estuaries likely experienced periods of hypersalinty, hypoxia, and anoxia. Such variable conditions would have affected species composition and the vigor of aquatic organisms, influencing growth rates, behavior, reproductive success, and survival. The response of aquatic organisms to low dissolved oxygen depends on the intensity of hypoxia, duration of exposure, and the periodicity and frequency of exposure. A range of physiological and behavioral adaptations would have allowed many resident organisms to deal with temporary periods of low oxygen availability and thus persist in the face of naturally variable conditions. Seasonal, annual, and longer-term dynamics also likely resulted in periodic patterns of species extirpation and colonization.

The idea of "water quality" must be viewed through the lens of species life history requirements across the full range of native taxa. Some prominent contemporary water quality issues, such as concerns about odor and aesthetics, have little relevance for understanding natural conditions. Furthermore, the conditions that give rise to contemporary eutrophication were not prevalent historically. In southern California systems, nitrogen and carbon are likely the limiting factors. Contemporary eutrophication is driven by nitrogen loading that occurs primarily during the summer (non-storm months), when the estuaries are subject to long exposure to sunlight, may have restricted tidal access, and may be stratified, allowing hypoxic or anoxic conditions to develop. These watershed loadings were mostly absent in the historical timeframe, and thus the probability of eutrophic conditions forming was likely much lower than today, even for periods when lagoons were closed to tides.

July 1769 and Los Peñasquitos Creek in June 1897; see Jepson 1898 and Crespí and Brown 2001). Even the San Dieguito River was recorded to run intermittently through broad alluvial areas such as the San Pasqual and San Bernardo valleys (e.g., Hall 1888).

Despite the apparent scarcity of summer surface water, groundwater levels in the valleys near the coast were described to be relatively high in the 19th and early 20th centuries prior to substantial extraction activities. Early descriptions of the San Diego region reported groundwater only a few feet from the surface, even in the dry season when streams supported little or no surface flow (e.g., *Daily Alta California* 1864, Van Dyke 1887, Holmes and Pendleton 1918, Ellis and Lee 1919): in August 1856, Hayes (1929) remarked that though no flow was visible in the Santa Margarita River, "water can be had anywhere by digging a few feet." The historical presence of extensive freshwater/brackish wetlands at the upslope margins of each lagoon, formed in high-groundwater areas where creeks spread into wetland complexes above each estuary, reflects the presence of at least some diffuse perennial freshwater inputs to each lagoon in the form of groundwater and surface runoff.

Numerous small wells were dug in stream beds across the region to take advantage of the presence of shallow groundwater, including in the San Dieguito River bed (Bronson 1968). This practice is memorialized in the name for Batiquitos Lagoon, mentioned by Spanish explorer Font in 1776. In northwestern Mexico, a *batequi* or *bategui* (derived in turn from the Yaqui *bate'ekim*) is a small well dug in the dry, sandy bed of a stream near the coast (Aschmann 1966, Gudde and Bright 1998, Des Lauriers and García-Des Lauriers 2006).

Sediment Inputs

Sediment accumulation in coastal systems is a natural process that allows shallow estuaries to keep pace with sea level rise. Sediment is derived locally (e.g., organic sediment from the marsh) as well from adjacent watershed and marine sources (e.g., mineral sediment transported during storm events). An imbalance in sea level rise and estuarine sedimentation rates can result in habitat type conversion. For example, sea level rise exceeding local sedimentation can result in the transformation of marsh to mud flat. Conversely, sedimentation exceeding sea level rise results in the eventual transformation of marsh to upland or non-estuarine habitat.

Estimates of historical sediment inputs to the lagoons are derived mostly from core data. Sediment cores are an important tool for deciphering the historical interplay between sea level rise and sedimentation rates. Though core data for the lagoons under investigation here are somewhat limited, the cores that have been analyzed for sedimentation rates over the past few centuries provide vital information for understanding the change in lagoon dynamics and ecosystem functioning since Euro-American settlement in the region.

While this report documents the characteristics of northern San Diego County lagoons circa 1800s, changes in the rates of sedimentation and sea level rise over time have altered lagoon character. The deep embayments present in North County during the late Pleistocene and early Holocene filled with sediment over thousands of years to become the shallow, intermittently tidal coastal lagoons documented by 18th and 19th

The Tia Juana, Sweetwater, San Diego, San Bernardo, San Luis Rey, and Santa Margarita rivers, and several smaller streams... are nearly all drying the Summer months for several miles from their outlets.

—GUNN 1886

There is not a stream in the country large enough and constant enough to drive a grist mill three months in the year; nor is there a stream that runs to the ocean throughout the year. The Santa Margarita, San Luis Rey, San Pasqual and San Diego rivers, all sink, except during very rainy seasons. Where, however, they pass through canons, there water is found at all times, while their beds in intervening plains are dry. ... the water percolates through the sand, and may be found at a depth of three or four feet.

—DAILY ALTA CALIFORNIA 1864

WHAT WERE NORTH COUNTY LAGOONS LIKE BEFORE THE WRITTEN RECORD?

Because an imbalance between the rates of sediment input and sea level rise can dramatically alter lagoon character, understanding changes in these rates provides important context for interpreting information about historical ecological patterns. Although there are accounts of the 18th century and early 19th century physical and ecological conditions for North County lagoons that note the presence of estuaries and salt flats (e.g., Duhaut-Cilly [1827]1997, Bliss 1846-7, Costansó and Browning 1992, Crespí and Brown 2001), they are by no means comprehensive in providing a detailed picture of historical lagoon conditions. As a result, it cannot be assumed that mid- and late 19th century records are representative of 18th century, pre-Euro-American settlement conditions with regard to average lagoon bed elevation, inlet closure dynamics, habitat conditions, and ecological functioning. Is it possible that lagoons were deeper, subtidal features immediately prior to Euro-American settlement – filling with fluvial sediment only following 19th century changes in land use practices – rather than the shallow, salt flat-dominated systems documented by historical observers?

To address this question, we assessed likely 18th century lagoon conditions using historical sedimentation rate and sea level rise estimates (see graph at right). We began by projecting the approximate elevation of salt flats relative to Mean Higher High Water (MHHW) from 1887-9 (the date of the detailed, comprehensive T-sheet surveys and descriptive reports) back to 1769 (the date of the first land-based European expedition in the region). Sea level rise rates were set at 1.5 mm/yr for the post Industrial Revolution time period (1850-1890, from Flick et al. 1999) and 0.75 mm/yr for 1769-1850 (from Inman 1983). Using historical sedimentation rates from Los Peñasquitos Lagoon, we developed an upper and lower estimate for the 1769 bed elevation. The upper estimate was derived using a pre-Euro-American settlement sedimentation rate of 1 mm/yr (Mudie and Byrne 1980) while the lower estimate was derived using a 1820-1840 sedimentation rate of 3.8 mm/yr (Cole and Wahl 2000). Both rates came from cores taken at locations that were high marsh at the time of sampling (bed elevation ~MHHW) and in areas mapped as salt flat in our synthesis mapping (the Mudie and Byrne core is on the margin of salt flat and salt marsh). While general, this analysis suggests that salt flat areas that were at about Mean High Water (MHW) in the late 19th century were at a similar elevation in the late 18th century. This finding implies that, in general, these lagoons were relatively shallow features in the late 18th century and did not experience a rapid transition from subtidal to intertidal in the period following Euro-American settlement.

To further assess the probability of rapid sedimentation altering lagoon character between the mid-18th century and late 19th century, we conducted a hypothetical analysis to estimate the sediment accretion rates that would be needed to convert a subtidal basin (average bed elevation between Mean Lower Low Water [MLLW] and ~1 meter below MLLW) to a shallow salt flat (average bed elevation ~MHW) during that period. Our analysis suggests that such a conversion would have required sustained average annual sedimentation rate of approximately 13 mm/yr to 21 mm/yr. These rates are quite high, greater than mid-20th century rates documented after major increases in watershed erosion following large-scale urban development. For example, the sediment accumulation rates at Los Peñasquitos and Mission Bay were estimated to be ~10 mm/yr during the development boom of the 1950s-1970s (Mudie and Byrne 1980), and rates of 7-12 mm/yr in the Tijuana River estuary were noted as the "high end of values measured in other coastal wetlands" (Weis et al. 2001).

From these data, we infer that sustained rates of 13-21 mm/yr for more than a century prior to major urban or agricultural development have no clear mechanism, and barring an acute disruption (e.g., a massive seismic event influencing bed elevation or sediment accretion rate) are not realistic. The earthquake of 1812, which was reported to cause widespread damage in other parts of southern California, was not reported to have damaged areas in coastal San Diego County (Agnew et al. 1979). In addition, we could find no evidence that subsequent major earthquakes that were felt in the region during the time period had an impact on bed elevations or watershed sediment supply (and subsequent lagoon sedimentation rates). Even if such a seismic event did occur, it is unlikely that it would have caused the same degree of sediment accretion or bed elevation change at all of the lagoons.

Similarly, the magnitude of impact on sediment accretion rates from 18th and 19th century land uses is not known. Grazing – first associated with the missions and pueblo of San Diego, and later with Mexican ranchos and American settlers – was the main pre-1880s land use; along with small-scale agricultural development, it may have altered sediment delivery to the lagoons. However, cattle densities were relatively low for most of the 18th and 19th centuries and were not uniform across watersheds, and thus grazing is unlikely to have affected all of the lagoons in the same way. Though significant impacts from grazing and agriculture cannot be discounted during this period, we believe they were unlikely to have caused such widespread, profound changes across all lagoons during this time period given the scale of these activities (see Chapter 3 for more detail on grazing and other land use impacts).

Of course, this analysis is not without limitations. In particular, the limited availability of published sediment accretion rates based on core data for these lagoons restricts certainty in our findings, since rates depend heavily on core location and would have varied across lagoons in time and space. However, the weight of evidence from sediment coring data combined with conservative sea level rise estimates suggests that the salt flats observed in North County estuaries in the late 19th century were also salt flats or other wetland habitat types in the late 18th century, prior to Euro-American settlement.

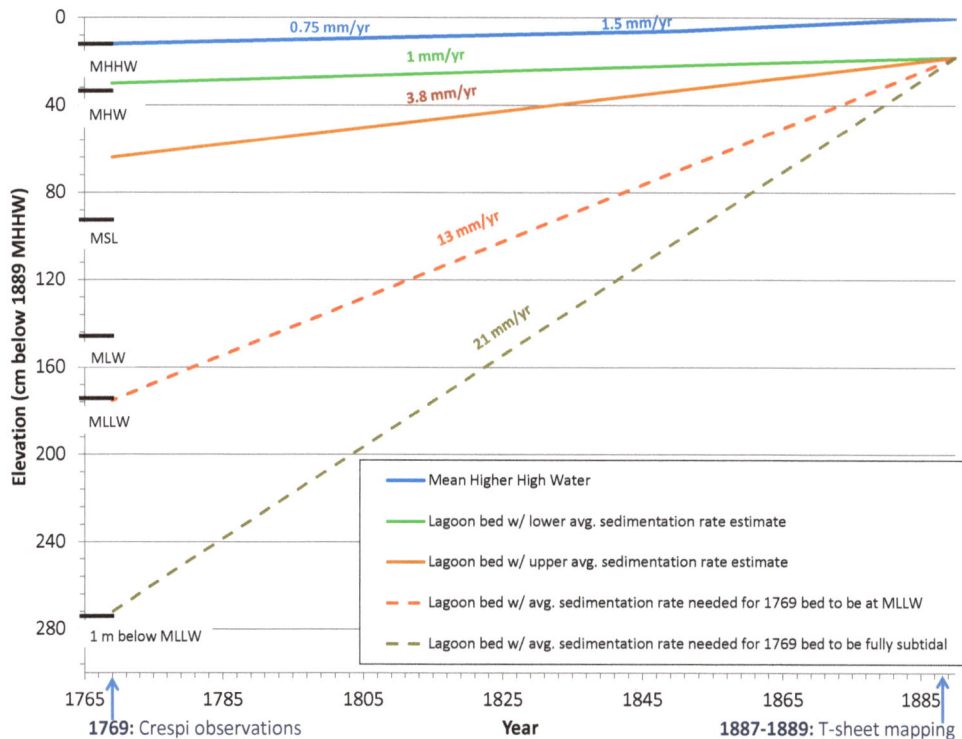

Calculated increase in MHHW and lagoon bed elevation at ~MHW for North County lagoons from 1769 to 1889. Solid lines represent "hindcast" bed elevation projections from 1889 back in time based on sediment core data and estimated historical local and regional sea level rise. Dashed lines represent hypothetical "hindcast" bed elevation projections associated with high sedimentation rates that would have had to exist for the 1769 bed elevation to be indicative of an open lagoon system.

century observers (Masters and Aiello 2007; see page 36). Published sediment core data for Los Peñasquitos Lagoon suggest a pre-Euro-American sedimentation rate of ~0.5 mm/yr (Cole and Wahl 2000) to ~1 mm/yr (Mudie and Byrne 1980). (In this discussion sediment accumulation rates are provided in the metric system to reflect the dominant way of presenting these figures in the literature; 1 mm=0.04 in.) These numbers are generally consistent with average pre-settlement rates documented for other northern San Diego County lagoons ranging from 0.4-1.6 mm/yr (see Mudie and Byrne 1980, Goodwin et al. 1992, Masters and Gallegos 1997) and are broadly comparable to estimates of late Holocene sea level rise rates of 0.5-1 mm/yr (Masters and Aiello 2007). It is important to note that these rates are averages only, and would have varied significantly both over time (with variations in land use and climate) and spatially (across estuaries and in different parts of each system).

By the 20th century, high sediment accretion rates induced by large-scale changes in land use were documented in many lagoons. Land conversion to agriculture, grazing, and urban uses resulted in an increased sediment supply to the lagoons, while bridges and berms built across the lagoon mouths restricted flow and may have increased sediment retention. The loss of freshwater/brackish transitional wetlands upstream of the lagoons, which historically buffered sediment transport to the lagoons, would have also increased sedimentation rates. The effects of dams constructed in the contributing watersheds during the 20th century on lagoon infilling, such as Lake Hodges (1918-19) and San Marcos (1952), are less clear: dams may have decreased sedimentation rates by trapping sediment and reducing supply delivered to the coast, but they may have also increased sediment retained in the lagoons by increasing in-channel scour below the dams and reducing the frequency and magnitude of lagoon-scouring floods.

The documented 20th century post-settlement sedimentation rates are in some cases many times higher than average pre-settlement rates and exceed rates of sea level rise (2.07 mm/yr from 1924-2006 at the La Jolla tide gage). Los Peñasquitos Lagoon, for example, experienced average sedimentation rates of about ~4-5 mm/yr and as high as 10 mm/yr during the 20th century (Mudie and Byrne 1980, Cole and Wahl 2000). These figures are consistent with high 20th century rates documented for other southern California systems, including San Elijo Lagoon (4 mm/yr or less; Goodwin et al. 1992, Laton et al. 2002), the Tijuana River (7-12 mm/yr from 1963-1998; Weis et al. 2001), and Mission Bay (5.9-10 mm/yr from 1910-1978; Mudie and Byrne 1980). These elevated accumulation rates undoubtedly impacted lagoon characteristics, including potential tidal prism volume, inlet closure dynamics, and ecological patterns and functions.

Inlet Dynamics

The historical dynamics of lagoon opening and closure are challenging to reconstruct given the variability between systems and fluctuations across, tidal cycles, seasons, and years. However, the historical data collected in this study do provide qualitative insight into lagoon inlet dynamics over time. In this section, we use three complimentary lines of evidence to create a more robust picture of inlet closure dynamics in the recent past. First, we synthesize early records of inlet condition from the 19th and early 20th centuries. Second, we estimate historical tidal prism volumes associated with the lagoons. Finally, pollen cores and microfossil records provide information about the extent of tidal influence during the historical era.

Direct Observations

METHODOLOGY Inlet condition data was compiled for all six lagoons from each pre-1940 historical source in which inlet condition could be unambiguously discerned. A total of 72 maps, photographs (including the 1928 historical aerial photomosaic), postcards, and textual records were analyzed, representing 128 depictions of inlet condition across the six systems (as some sources contained information for multiple lagoons). Inlet condition as described or depicted by each source was classified as either "open" or "closed"; sources in which inlet condition was ambiguous were not included in the analysis. Our definition of "open" included any state in which ocean water could enter the estuary, even if circulation was limited to a narrow tidal range (i.e., all conditions with closure anywhere in the intertidal or subtidal zone). We defined a "closed" system as closed to tidal influence at high tide, corresponding to the conditions described in Jacobs et al. (2010) as "dune dammed," "perched," and "closed." The date and season of each source was also recorded where available.

Inlet condition was classified as closed in maps showing a lagoon separated from the coast or a two-dimensional channel separated from the ocean by a beach or solid coastline. Inlet condition was classified as open in maps showing a lagoon or two-dimensional channel connected to the ocean (i.e., with a broken coastline; see page 178 for example of maps depicting inlet condition). On photographs, lagoons were classified as open if there was a connection apparent between the lagoon and ocean. Textual sources that referenced one or more specific dates or time periods were counted; observations that made general reference to typical inlet conditions but did not describe inlet conditions during a specific time period were not included. In some cases, a single quote provided multiple data points if it specified inlet conditions during multiple time periods (table 11.1, page 178).

A cut-off date of 1940 was chosen for sources used in the analysis. While both direct and indirect manipulation of the lagoon inlets had already occurred by this time, 1940 is the first year in which permanent stabilization of any of the lagoon inlets is documented (installation of a weir at Buena Vista Lagoon); this date therefore captures the maximum number of historical sources while omitting sources that post-date permanent inlet modification. We also performed a subset of the analysis using a cut-off date of 1912 (the year in which construction began on the Coast Highway) to assess whether the use of an earlier cut-off date altered the results substantially.

This is a relatively coarse analysis, and several caveats should be mentioned. First, though for simplicity we used a binary open/closed classification, inlet condition is in reality much more complex (see Jacobs et al. 2010). For example, the T-sheet shows San Dieguito Lagoon with an open mouth and was thus counted as a depiction of "open" conditions, even though the position of the Mean Lower Low Water line suggests that the lagoon would have often been disconnected from the ocean at low tide (Rodgers and Nelson 1889; cf. Jacobs et al. 2010). A few observations capture these subtleties in inlet condition, offering more nuanced insights into system dynamics not captured in this analysis (e.g., Duhaut-Cilly [1827]1997; Osgood 1881a,b; Rodgers and McGrath 1887-8a,b; Rodgers and Nelson 1889).

Second, since the season in which an observation was made is unknown for the majority of sources and those observations for which seasonality is known are not evenly distributed throughout the year, this was not factored into this coarse analysis. Perhaps most importantly, it must be emphasized that these data represent the *depiction* of inlet condition at a given point in time, which may be distinct from the *actual* inlet condition at that time. As a result of these limitations, these data are meaningful for understanding general trends in historical inlet dynamics, but cannot be translated into lagoon closure frequencies.

Table 11.1. Selected quotes from historical documents pertaining to inlet condition at North County lagoons.

Lagoon	Date	Quote	Source
San Dieguito	November 1874	"To mouth of San Diegito River, where it empties into Pacific Ocean."	Goldsworthy 1874a
Batiquitos	February 1875	"This marsh is shut out from the Ocean by a wall of sand and cobble stones along the beach."	Wheeler 1874-5
Los Peñasquitos	1883	"A sort of natural breakwater, through which now the sea can pass only at the northern side, in the spring tides and storms, to mingle with the fresh water of the creek."	Wilson 1883
Batiquitos	spring 1884	"Batiquitos Lagoon waters began flowing to the ocean after the spring flood of 1884."	Ball 1976
Buena Vista, Agua Hedionda, Batiquitos	1887-8	"Protected now [summer] from the break of sea waves by dykes of sand or shingle. During the wet season, they are overflowed by fresh water and storm waves break over the front dykes mentioned, when the area for a mile inland from the sea forms a shallow lagoon."	Rodgers 1887-8b
Batiquitos, San Elijo, San Dieguito	1887-8	"There are shingle levees in front of San Marcos...and San Alejo and San Dieguito protecting them from the free entrance of breaking waves of the ocean: these levees are cut at their north ends and there permit the ebb and flow of the higher tides through narrow openings."	Rodgers 1887-8a
San Dieguito	1888	"From San Pascual valley, the river next passes through Bernardo valley, enters an open cañon a few miles from the coast, and emerges into a low, narrow valley through which it runs sluggishly to the sea, joining a tidal estuary at its mouth."	Hall 1888
San Dieguito	1913	"The mouth of the San Dieguito is closed in summer...forming a sand bar."	Post 1913
Los Peñasquitos	1933-34	"The mouth of Soledad Valley was completely closed by a sand bar at the time of the photographs and also at the time of the plane table survey. Water stands in the channels back of the sand bar; and overflows onto the mud flats during the accumulation of rain in the winter. Just as soon as this water area attains a level sufficient to break the sand bar a channel to the sea is rapidly formed and the area is drained of its fresh water."	Knox 1934-5a
Batiquitos	1934	"The mouth of Batiquitos Lagoon was closed by a sand bar."	Knox 1934-5b
San Elijo	1934	"The mouth of San Elijo Lagoon was open to the sea when the photographs were secured, but completely closed by a sand bar when the plane table survey was run...During the dry season the mouth of the lagoon becomes entirely closed by tide action...during the rainy season the mouth is again opened and it becomes tidal until this action is repeated."	Knox 1934-5b
Buena Vista, Agua Hedionda	1934	"A sand bar completely closed the mouths of all drainages [on sheet #T-5412] except Agua Hedionda Creek, which was open at the time of the photographs and also at the time of the planetable survey of the coast-line."	Knox 1934b

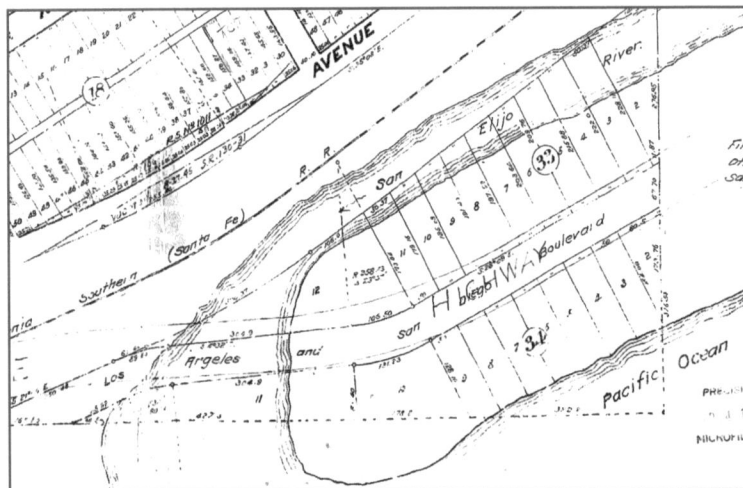

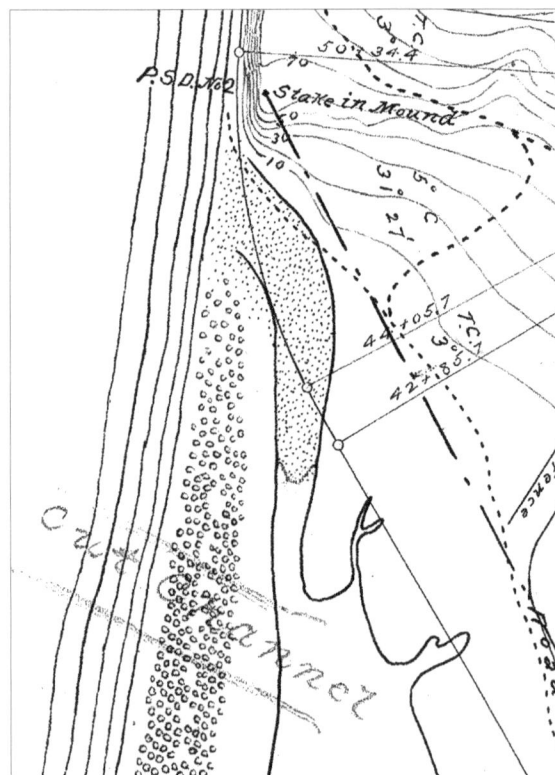

Maps such as those pictured here provided evidence for intermittent opening and closure at each of the six lagoons throughout the 19th and early 20th centuries. **(top)** This map of Cardiff from June 1910 depicts the "San Elijo River" with an open connection to the ocean. **(right)** A California Southern Railroad survey map from 1888 shows a stippled pattern at the mouth of Los Peñasquitos Lagoon, indicating that it was blocked by sand or cobble. (top: Rumsey & King 1910, courtesy of San Diego Cartographic Services; right: Unknown 1888b, courtesy of California State Railroad Museum)

RESULTS Of the 128 pre-1940 observations of inlet condition, 70 depicted an open inlet and 58 depicted a closed inlet (table 11.2, pages 180-181). (Again, these numbers do not imply the percentage of time the lagoons spent in a given state.) The number of historical observations for each lagoon ranged from 17 to 26.

While closure frequencies for a given lagoon cannot be determined through this analysis, it is important to note that all six lagoons were depicted as both open and closed by numerous historical sources. No lagoon was shown as always open or always closed; on the contrary, each lagoon was depicted in both states by between five and twenty independent historical sources. Similar patterns are also evident when the dataset is limited to only pre-1912 sources.

In addition, the frequency with which each lagoon inlet was depicted as open or closed varied across lagoons, from Buena Vista Lagoon (shown as closed by 13 sources, and open by 6) to San Dieguito Lagoon (shown as closed by 6 sources, and open by 20). While these variations in and of themselves are not sufficient to demonstrate real variations in inlet dynamics across lagoons, they are suggestive of patterns that appear to be corroborated by other analyses based on our historical ecological reconstruction.

Tidal Prism Analysis

In addition to direct early observations, estimates of historical tidal prism volume based on our historical mapping shed light on lagoon closure dynamics in the 19th century. Historical tidal prism volumes have been estimated by previous researchers for two North County lagoons: Batiquitos and San Dieguito. San Dieguito Lagoon was estimated to have a historical (1889) potential mean tidal prism volume of 24,000,000 ft^3, while Batiquitos Lagoon's historical (1850) potential mean tidal prism volume was calculated to be approximately 60,000,000 ft^3 (Coats et al. 1989). These volumes have been used to infer that both lagoons were fully tidal in the mid- to late 19th century (Coats et al. 1989).

However, these tidal prism estimates are problematic. The methods used to determine the San Dieguito Lagoon historical tidal prism volume, which is presumably based on the earliest T-sheet given the volume date, were not found. The Batiquitos Lagoon volume originates from a 1978 report investigating the lagoon's historical hydrology which assumed that the lagoon was subtidal (with its boundary at Mean Sea Level) and had a high sustained sedimentation rate throughout the mid- to late 19th century that resulted in a loss of about half of the lagoon volume from 1850 to 1978 (Gayman 1978a, Phillips et al. 1978). Since these assumptions – in particular regarding the historical lagoon bed elevation – are not supported by the more extensive reconstruction of historical ecological patterns and lagoon elevations based on the archival research presented here, we calculated an estimated historical tidal prism volume for each lagoon based on our revised mapping.

METHODS To estimate historical tidal prism volume, we combined our mapping of historical ecological patterns with contemporary tidal datum information for San Diego (NOAA gage 9410170). We assumed that the approximate difference between current and historical tidal datums is similar (i.e., the difference between Mean Lower Low Water [MLLW] and Mean Higher High Water [MHHW] is more or less the same now as it was over a century ago). We then assigned a representative surface elevation relative to historical MHHW for each habitat type based on our understanding of vegetation zonation within southern California lagoon systems (table 11.3, page 184). The emergent salt marsh elevation was set at MHHW based on the distribution of marsh plain and high marsh plant species in San Diego County (Sullivan 2001). Average open water/mud flat elevation was set at

Year	Season	Source	Buena Vista	Agua Hedionda	Batiqui-tos	San Elijo	San Dieguito	Los Peña-squitos
1827	June	Duhaut-Cilly [1827]1997					O	
ca. 1840	Unknown	USDC ca. 1840b		C	C			
1853	Unknown	Foster 1853	C					
1861	Unknown	Williamson 1861						O
1869	March	Hoffman Bros. 1869				O		
1872	Unknown	Wheeler et al. 1872	C	C				
1874	November	U.S. Surveyor General's Office 1876; Goldsworthy 1874a					O	
1875	February	U.S. Surveyor General's Office 1875, 1883, 1890; Wheeler 1874-5			C			
1881	January	Unknown 1881	O	O	O	O	C	
1881	February	Osgood 1881a	C	C	C	C	O	
1881	February	Osgood 1881b	C	C	C	O	O	
1881	March	Osgood 1881c	O	O	O			
ca. 1881	Unknown	Unknown ca. 1881a	O	O	O	O	O	
ca. 1881	Unknown	Unknown ca. 1881b	O	O	O	O	O	
1883	Unknown	Wilson 1883						C
1883	Unknown	Unknown 1883 in Gayman 1978b			O			
1883	Unknown	Wilson 1883						O
1884	Unknown	Ball 1976			O			
1887	January	Couts 1887	C					
1887-8	Wet season	Rodgers 1887-8b	O	O	O			
1887-8	Dry season	Rodgers and McGrath 1887-8a; Rodgers 1887-8b	C	C	C			
1887-8	Unknown	Rodgers 1887-8a					O	
1887-8	Unknown	Rodgers and McGrath 1887-8b; Rodgers 1887-8a				O		
1888	June	Unknown 1888b						C
1888	June	Unknown 1888c						C
1888	Unknown	Hall 1888					O	
1889	Jan-Feb	Mansfield 1889				O		C
1889	Unknown	Beasley and Schuyler 1889					O	
1889	Unknown	Rodgers and Nelson 1889; Rodgers 1889					O	O
1891-8	Unknown	USGS 1891, [1891]1898	C	C	C	C		
1898	October	Fifth Road District Survey Map 1898 in Gayman 1978b			C			
ca. 1900	Unknown	Unknown ca. 1900						C
1903	Unknown	USGS 1903					O	O
ca. 1907-14	Unknown	Unknown ca. 1907-14						C
1910	June	Rumsey & King 1910				O		
1913	February	South Coast Land Co. 1913					C	

Table 11.2. Inlet condition as depicted and described in historical sources by lagoon, 1827-1940. Sources that provided multiple data points for different time periods appear in the table more than once. Data are insufficient to draw conclusions about closure frequency or duration for any individual lagoon or year: for example, a year when all lagoons are shown as open could reflect particularly wet conditions

Year	Season	Source	Buena Vista	Agua Hedionda	Batiquitos	San Elijo	San Dieguito	Los Peñasquitos
1913	Summer	Post 1913					C	
1913	Unknown	Unknown 1913						C
1913	Unknown	Wood 1913	C	C	C	C		
1914	Unknown	Unknown 1914 in Hawthorne 2003						O
ca. 1915	Unknown	Unknown ca. 1915b						O
ca. 1915	Unknown	Schwartz and Ewing ca. 1915						O
1916	August	Harris and Simmons 1916					O	
1919	Unknown	Butler 1919	C	C	C			
1919	Unknown	Ellis and Lee 1919	C	C	C	C	O	O
1920	Unknown	Rodney Stokes Co. 1920					O	
1922	Unknown	Kelly in Harmon 1967		O				
ca. 1925	Unknown	Unknown ca. 1925a						C
1926	Unknown	California Highway Commission 1926 in Gayman 1978b			O			
1927	February	Unknown 1927					O	
1927	Unknown	Kelly in Harmon 1967		O				
1928-29	Nov-Mar	San Diego County 1928	C	O	C	O	O	C
1928-29	Nov-Mar	San Diego County 1928 (ortho1928_45b1.img)					C	
1930	Unknown	USGS 1930					O	O
1930	Unknown	States Publishing Co., Ltd. 1930	O	O	O	O	O	O
1932	February	Unknown 1932 in Elwany et al. 1995					O	
ca. 1932	Unknown	Unknown ca. 1932						C
1933-34	December	Knox 1933-4; Knox 1934-5a					C	
1933-34	Dec-Jan	Knox 1934-5a						O
1933-34	Dec-Jan	Knox 1934-5a						C
1933-34	Unknown	Knox 1933-4; Knox 1934-5a						C
1933-34	Unknown	Knox 1934-5a					C	
1934	January	Knox 1934a; Knox 1934b	C					
1934	January	Knox 1934d; Knox 1934-5b			C			
1934	January	Knox 1934-5b				O		
1934	January	Knox 1934b		O				
1934	Mar-May	Sipe 1934		O				
1934	Unknown	Knox 1934a; Knox 1934b		O				
1934	Unknown	Knox 1934d; Knox 1934-5b					C	
1934	Unknown	Knox 1934-5b			C			
1937	June	Unknown 1937 in Elwany et al. 1995					O	
1938	Unknown	Pirazzini 1938	C					
1939	April	Unknown 1939				O		

or surveyor bias. However, taken as a whole these data show that each lagoon was both intermittently closed and open to the ocean throughout this time period.

CASE STUDY: SAN DIEGUITO LAGOON INLET DYNAMICS

San Dieguito Lagoon's ~350 square mile watershed is by far the largest watershed of any of the six lagoons, about four times larger than the next largest watershed (Los Peñasquitos Lagoon) and over 15 times the size of the smallest (Buena Vista Lagoon; see graph on page 37). It also had the largest potential mean tidal prism volume of any of the six lagoons (see table 11.4). These characteristics, coupled with 26 historical (pre-1940) observations of inlet condition (see table 11.2), suggest that San Dieguito Lagoon was frequently connected to the ocean during the 19th century.

During periods of heavy rainfall, large flows from the San Dieguito River would have breached the beach berm, opening the lagoon inlet. Once the beach barrier was overtopped, the lagoon's tidal prism may have been sufficient to keep the lagoon open for extended periods (Elwany et al. 1998). The summer (May-July) 1889 T-sheet survey (Rodgers and Nelson 1889) shows the San Dieguito River mouth open, though the location of the MLLW line shows it would have been cut off from tidal influence at low tide. Goldsworthy (1874a; November 26) noted that the mouth was open, as did Duhaut-Cilly (June 1827[1997]; see below). Ellis and Lee (1919), in a reconnaissance predating the construction of Hodges Dam, reported that "only the Santa Margarita, the San Dieguito, and the Soledad [Los Peñasquitos] are able to keep narrow channels open through the beach deposits." One of the earliest railroad surveys (February 1881) also shows San Dieguito as unambiguously open, in contrast to the four lagoons to the north which are shown more ambiguously with a solid, unbroken coastline separating the lagoon from the ocean, potentially indicating lagoon closure (Osgood 1881a).

Despite having a much larger watershed than the other lagoons, however, even San Dieguito Lagoon did not maintain a permanently tidal connection to the ocean. During the dry season, significant flows to the lagoon often ceased (see page 124), and wave action caused sand accumulation in the channel. Some early descriptions allude to the presence of a bar across the mouth, which muted flow even when the mouth was open. Rodgers (1887-8a) wrote that the San Dieguito River mouth would "permit the ebb and flow of the higher tides through narrow openings which connect with shallow tidal sloughs," suggesting that only relatively high tides would overtop the bar. In June 1827, French sea captain Auguste Duhaut-Cilly recalled the challenge of crossing the river during heavy flows above the sand bar, which muted but did not block the tides:

> We came to a large stream called the Estero de San Dieguito, which rushed into the sea with great force, forming a turbulent bar where it met with the waves. With no hesitation the Californians boldly rode into the torrent, and I, under pain of being left behind, had perforce to follow them, and it was not without difficulty that we reached the other side. Although we took the precaution of heading our horses up into the current, we were carried downstream and came out well below our starting point and quite close to the bar, breaking only two fathoms away and almost over our heads like a fearsome arch above us. When all had crossed without mishap, we took up again our fast run along the beach for another seven leagues. (Duhaut-Cilly [1827]1997)

Others observed complete periodic closure of the mouth during the dry season. For example, a fisheries record from 1921 noted an "inclosed salt water lagoon" at the mouth of the river (Hubbs 1921), and Post (1913) stated that "the mouth of the San Dieguito is closed in summer by tide current, wave and wind action, forming a sand bar. Behind the bar is formed a lagoon" (see map at right). In these cases, subsequent freshwater inputs would be trapped behind the sand barrier until waves or flood waters overtopped the barrier and cut a new channel through the beach. Post further describes these summer conditions:

> The lagoon probably stands in summer at elevation 2.5 feet or slightly above mean tide. When water is flowing down the San Dieguito this height will rise, until at about 5 to 6 feet elevation, it will force a channel through the bar, when for a brief period the lagoon may ebb and flow between nearly 0 and 4.8 ft. elevations. The bar will soon reform and the general and practical condition may be considered to be an elevation of 2.4 ft to 4.8 ft. (Post 1913)

The mouth of the channel of this stream bed was completely closed by a sand bar at the time of the photographs [Dec 1933-Jan 1934] as well as at the time of the plane table survey. There is an extensive system of channels back of the mouth of this stream bed; and as is the case at the mouth of the Soledad Valley, this area is changeable in character, varying with the seasons.

—KNOX 1934-5A

By the mid-1900s, anthropogenic impacts – in particular, flow regulation on San Dieguito River from Hodges Dam, which was constructed in 1918 and impounds over 85% of the river's watershed – had altered inlet dynamics noticeably, restricting large flood events and producing more frequent closure conditions and a frequently dry tidal channel during the summer (Knox 1934c, Purer 1942). Researchers reported that the lagoon inlet remained largely closed from 1946 to 1977 (a relatively dry period) except for after a few large winter storms, unusually high tides, or artificial opening to drain the lagoon (San Diego Regional Water Quality Control Board 1967, Mudie et al. 1976, Elwany et al. 1998); Dailey et al. (1974) called the lagoon "permanently closed to the ocean." This trend reversed somewhat in the late 1970s and the inlet was open intermittently from 1978 to 1994 (a relatively wet period), though it was still only estimated to be open about one third of the time (Elwany et al. 1995, Elwany et al. 1998).

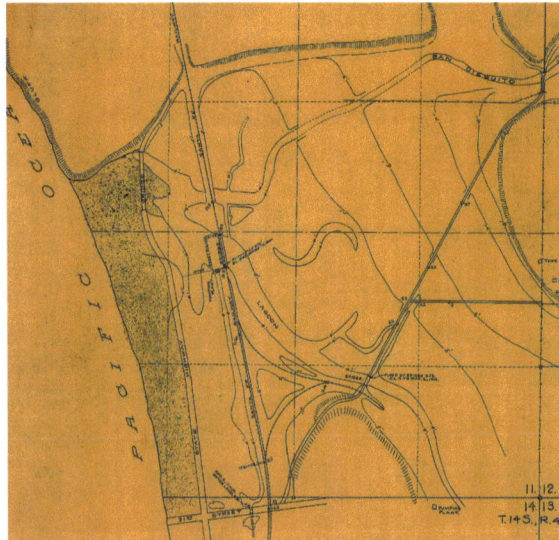

Map of the mouth of the San Dieguito River in February 1913. This was a winter with below-average rainfall, and the mouth is shown as closed. The lagoon periodically became isolated from the ocean during low-flow episodes, but waves or flood waters eventually re-established the tidal connection. (South Coast Land Co. 1913, courtesy of Holdings of Special Collections & Archives, UC Riverside)

This ca. 1954 oblique aerial photograph shows the closed San Dieguito River mouth, its dominant condition for many decades during the mid-1900s (in contrast with 19th century conditions). (Collection 87-26, courtesy of Scripps Institution of Oceanography Archives, UC San Diego)

Table 11.3. Relative elevation of each habitat type used for historical tidal prism volume estimations.

Habitat Type	Average Surface Elevation
Emergent Salt Marsh	MHHW
Salt Flat (Seasonally Flooded)	0.5 ft below MHHW
Channel (Open Water/Mud Flat)	MLW (4.8 ft below MHHW)

Mean Low Water (MLW), which was derived by calculating the maximum channel depth below MHHW at the lagoon mouth using hydraulic geometry relationships from Williams et al. (2002) and scaling down that value to represent a lagoon-wide average channel depth. Average salt flat elevation was assigned based on observations from Rodgers (1887-8b), who noted that the flats were "but little above the level of Mean High Water" (MHW) at Buena Vista, Agua Hedionda, and Batiquitos lagoons.

Historical potential mean tidal prism values were then estimated for each lagoon by combining the potential mean tidal prism (lagoon volume between MLW and MHW) calculated for each habitat type based on our historical synthesis mapping. While assessments of historical tidal prism volumes are clearly based on coarse assumptions about average ground surface elevations, they are useful in providing a general understanding of the range of historical tidal prisms across these six lagoons and a sense of how their historical potential tidal prism compares to previous calculations with a similar, if not greater, degree of uncertainty.

RESULTS Our estimates of historical potential mean tidal prism volumes ranged from 620,000 ft³ (Buena Vista Lagoon) to 6,100,000 ft³ (San Dieguito Lagoon; table 11.4 below). These figures are far lower than previous estimates: for Batiquitos Lagoon, the revised estimate is ~50 times smaller than the previous estimate from Gayman (1978a); for San Dieguito Lagoon the revised estimate is about four times smaller than the previous estimate from Coats et al. (1989). It is important to note that the previously reported value for Batiquitos Lagoon was derived by increasing the estimated historical lagoon volume by assuming very high historical sedimentation rates (~1-2 cm/yr for Batiquitos Lagoon; Phillips et al. 1978) derived from limited data sources. When used to determine historical lagoon bed elevations, these high rates yielded high historical lagoon depths and associated tidal prisms. In addition, the previous studies did not have the benefit of the historical site descriptions and other textual and spatial data sources used in the analysis presented here.

These tidal prism calculations have significant implications for the presumed historical lagoon closure frequencies and associated ecological functioning, as illustrated by the plot of tidal prism against wave power (Johnson 1973, Coats et al. 1989, Battalio et al. 2006; see facing page). While this graph addresses

Table 11.4. Estimated diurnal and mean potential tidal prism volumes based on historical synthesis mapping.

Lagoon	Est. diurnal tidal prism (ft³)	Est. mean tidal prism (ft³)	Previous est. diurnal tidal prism (ft³)	Previous est. mean tidal prism (ft³)
Buena Vista	5,800,000	620,000	–	–
Agua Hedionda	7,700,000	3,400,000	–	–
Batiquitos	12,000,000	1,300,000	90,000,000[1]	60,000,000[1]
San Elijo	12,000,000	5,200,000	–	–
San Dieguito	7,300,000	6,100,000	37,000,000[2]	24,000,000[2]
Los Peñasquitos	5,500,000	3,000,000	–	–

[1] from Gayman 1978a

[2] from Coats et al. 1989

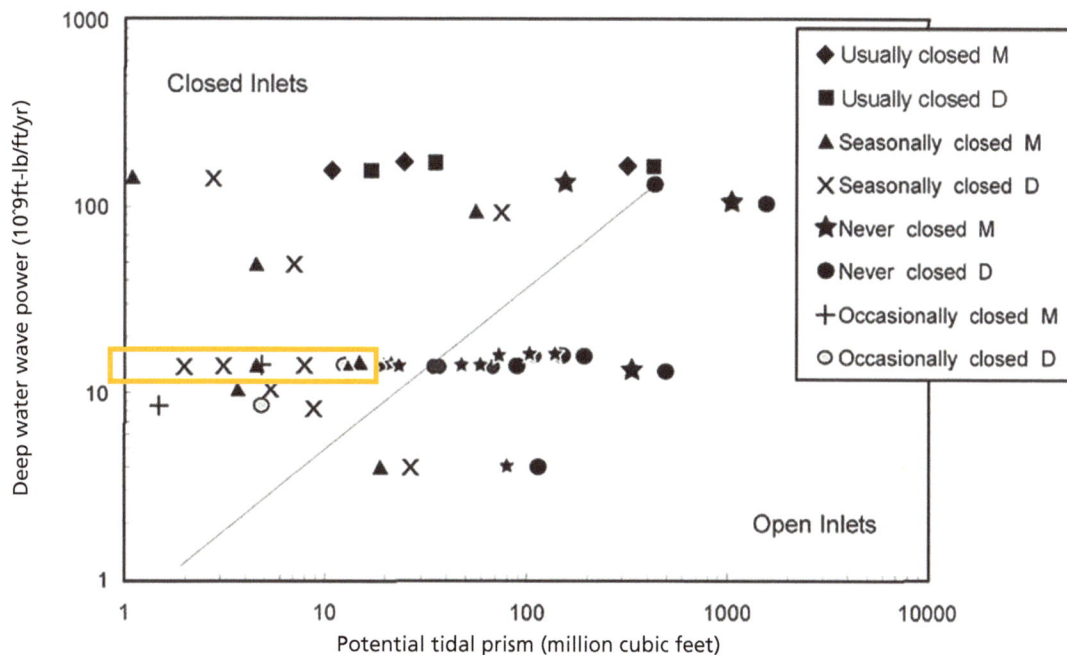

Power-based index of inlet closure for several California lagoons as a function of tidal prism (diurnal [D] and mean [M]) and wave power (from Battalio et al. 2006; the individual lagoons represented by circles, crosses, and stars on the graph are not specified by the authors). The approximate range of points for the six northern San Diego County lagoons based on our recalculations of mean and diurnal tidal prism volume are shown by the yellow box.

only a limited set of factors – it does not capture all elements relevant to understanding lagoon closure patterns (watershed dynamics in particular are not represented), for example, and the relationship may not hold for small lagoons (Prestegaard 1975) – it does provide a general picture of the tidal prism and wave power characteristics that generally support closed, frequently closed, and typically open inlet conditions for California lagoons. Combining this graph with the tidal prism estimates from this analysis suggests that the six lagoons closed intermittently, with the smaller lagoons likely being opened predominantly by large flood events.

Core Data

In addition to direct observations of inlet condition and tidal prism analysis based on ecological synthesis mapping, analysis of core data (in particular, pollen and microfossil records) from the lagoons provides insight into historical and prehistoric environmental conditions. Published analyses of core data for these systems pertain almost exclusively to Los Peñasquitos Lagoon, which represents the longest pollen record available for southern California (Mudie and Byrne 1980, Cole and Wahl 2000, Scott et al. 2011). Additional core data for other lagoons appears to be relatively limited and mostly unpublished (e.g., Phillips et al. 1978 for Batiquitos Lagoon and Pope 2004 for San Elijo Lagoon) and/or taken for non-paleoecological purposes (e.g., Caltrans boring logs collected in the 1960s along the I-5 corridor prior to highway construction).

Core data have been used to address a variety of paleoenvironmental research objectives, including reconstruction of paleoclimate and system evolution, vegetation communities, and sediment accretion rates (see page 174 for a discussion of sedimentation rates from core data). In a few cases,

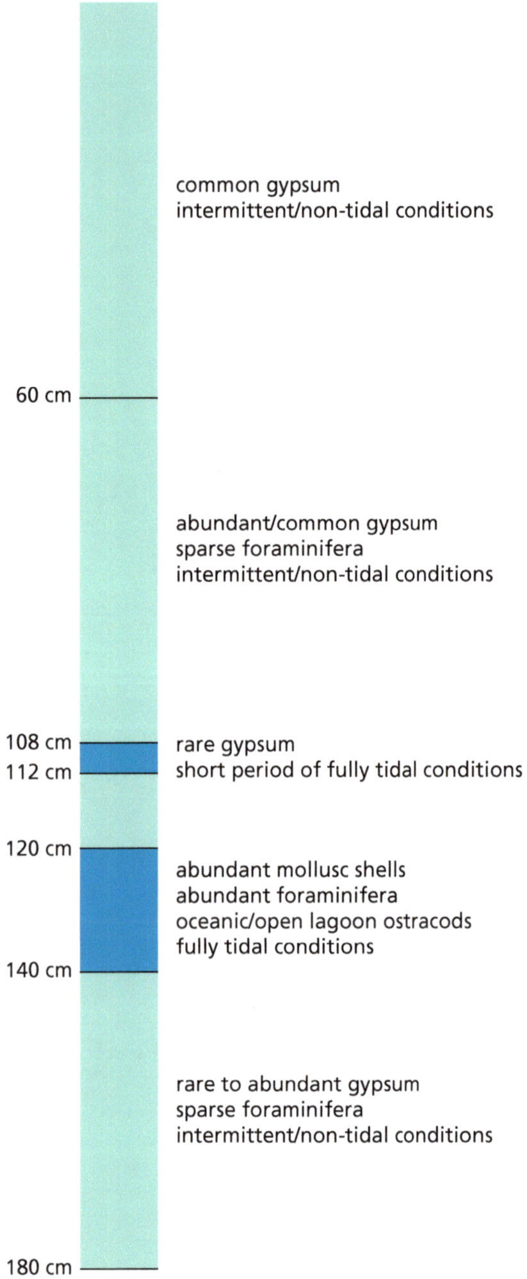
common gypsum
intermittent/non-tidal conditions
60 cm
abundant/common gypsum
sparse foraminifera
intermittent/non-tidal conditions
108 cm
112 cm
rare gypsum
short period of fully tidal conditions
120 cm
abundant mollusc shells
abundant foraminifera
oceanic/open lagoon ostracods
fully tidal conditions
140 cm
rare to abundant gypsum
sparse foraminifera
intermittent/non-tidal conditions
180 cm

Depiction of core Ba-7 DS-10, adapted from Phillips et al. (1978). Phillips et al. interpreted the degree of tidal exchange at different points in the core from the distribution and abundance of gypsum, mollusc shells, foraminifera and ostracods, as summarized here. Darker blue areas represent portions of the core that were interpreted by Phillips et al. as more marine-influenced, while ligher blue areas were interpreted as more intermittently tidal conditions.

researchers have also used microfossils (ostracods and foraminifera) to investigate the degree of tidal influence experienced by lagoons over time, often over a time span of many thousands of years (e.g., Pope 2004). These reconstructions, while instructive at a broad spatial and temporal scale, often do not interpret results at a resolution relevant to the historical era, and strata can be challenging to date precisely.

While there are no definitive reconstructions of tidal conditions with firm dates for any of the lagoons, findings from two studies are worth noting. Scott et al. (2011), drawing on cores sampled in the 1970s, reconstructed the general paleoecology of Los Peñasquitos Lagoon back to ~6,000 BP, showing transitions between mud flat, marsh, lagoon, and enclosed bay over the mid- to late Holocene. They conclude that the lagoon was deeper and more open to tidal influence in the mid-Holocene than today, and that the lagoon began to sediment in and become "semi-isolated from tidal inflow" prior to Mission-era land use activities. Phillips et al. (1978; see also Meyer 1980) used the microfossil record from a series of cores from Batiquitos Lagoon to demonstrate alternating continuously tidal and intermittently tidal conditions in the recent past. The authors found sparse fossil records in the upper portions of several cores, suggesting intermittent tidal flooding "mixed with periods of drying and/or fresh water inundation" as well as common or abundant gypsum through much of the core, indicating evaporative conditions.

Phillips et al. (1978) provide a chronology for core Ba-7 DS-10 (pictured at left) that suggests two periods of continuously tidal conditions bracketed by intermittently tidal conditions at Batiquitos Lagoon from ~1830-1978. Based on an average sedimentation rate of 11 mm/yr, they suggest that the two continuously tidal periods occurred between about 1850-1870 and about 1877-1881. However, this interpretation is based on a very high sustained sedimentation rate, particularly for the 19th century (see page 174). While this rate is generally consistent with rates estimated for Los Peñasquitos Lagoon during the mid-20th century (see Mudie and Byrne 1980, Cole and Wahl 2000), it is much higher than estimated 19th century post-settlement, grazing-era rates at Los Peñasquitos (average 3.5 mm/yr). It is not clear what the mechanism for such elevated rates at Batiquitos Lagoon would have been during the early- and mid-1800s. In addition, this interpretation is weakened by historical sources which describe Batiquitos Lagoon as closed to tidal influence during this time period (Wheeler 1874-5) or depict likely intermittent closure (e.g., USDC ca. 1840b, Osgood 1881a).

Though there are insufficient data to definitively determine the precise duration and timing of periods of tidal conditions at Batiquitos Lagoon, a few conclusions may be drawn using core chronologies based on lower sedimentation rates. For example, a more conservative estimate of ~1-2 mm/yr for the average pre-20th century sedimentation rate (see page 174) would shift the date of the base of the core to ca. 900-1400 AD, and the timing of the longer period of increased tidal influence several centuries earlier, to ca. 1260-1460 (1 mm/yr) or ca. 1585-1685 (2 mm/yr). This revised interpretation of the Batiquitos chronology from the Philips et al. (1978) cores is broadly consistent with climatological evidence: these dates coincide with portions of the Little Ice Age (ca. 1300-1850), when the California climate was generally cooler and wetter than it was during the late 19th and 20th centuries. The difference in climate may have resulted in increased runoff, more frequent opening of the lagoon inlet, and a more regular tidal connection during some periods (Roger Byrne, pers. comm.). In addition, radiocarbon dating of a core taken by Miller (1966) in the northeastern portion of Batiquitos Lagoon yielded a calibrated median date of 1570 for a layer containing abundant shell deposits. This likely represents the same shell deposit identified in the Philips et al. (1978) cores, further supporting the conclusion that the period of sustained marine influence at Batiquitos Lagoon occurred during the middle of the Little Ice Age (Roger Byrne, pers. comm.).

The limited available core data present a complicated and intriguing window into prehistoric and historic tidal dynamics for selected systems. However, for the most part the data are not sufficient to draw conclusions about the precise timing of environmental conditions. Importantly, the Batiquitos core data illustrate that lagoon evolution has not necessarily been strictly linear (i.e., deep embayments becoming shallow lagoons) over the mid- to late-Holocene. Instead, they suggest that periods of more continuous tidal influence would have alternated with periods of more intermittently tidal conditions, likely following paleoclimatic events such as megadroughts and megafloods (Masters and Aiello 2007). Additional core data may provide further resolution on the history of tidal dynamics in the lagoons.

Summary and Discussion

Inlets are dynamic features: opening and closing, scouring and filling, and changing location over time. The precise closure regime of these six lagoons cannot be reconstructed from these data, and of course would have been variable from lagoon to lagoon, year to year, and across seasons. Despite this uncertainty, multiple lines of evidence – direct historical observations, estimates of historical tidal prism volumes, and core data – suggest that northern San Diego County lagoons were intermittently closing and opening systems in the 1800s, including prior to the construction of the railroad. This is in contrast to previous assertions that many of these lagoons were "fully" or "continuously" tidal estuaries as late as the 1880s and that the intermittent conditions recorded by 19th and early 20th century observers were a function of anthropogenic impacts (e.g., Mudie et al. 1974, State Coastal Conservancy 1987, Marcus 1989, Merkel & Associates, Inc. 2009, Byrd n.d.).

This reassessment of historical closure dynamics is also supported by inference from historical ecological and geomorphic patterns documented across many lagoons. The extensive salt flats first documented by the Portolá Expedition of 1769 were positioned a little higher than MHW, and were occasionally tidally inundated when lagoon mouths were open (e.g., see Harmon 1967). For salt flats to crystallize salt and persist over the dry season at this elevation, they would have had to have been protected from tidal flooding by a berm. Low historical channel densities through the marsh plain in all six systems also provide evidence for intermittent closure. The northern three systems displayed relatively undeveloped channel networks, which is indicative of irregular tidal exchange (Phil Williams, pers. comm.). In contrast, the channel networks of the southern three systems appear to have been somewhat more developed, potentially representing evidence for differences in opening frequency and/or duration between the northern and southern lagoons (Josh Collins, pers. comm.). However, none of the systems supported channel densities comparable to those found in fully tidal systems (Coats et al. 1995).

These findings are generally corroborated by results from Jacobs et al. (2010), which predicted closure patterns for southern California estuaries based on coastal setting, exposure, watershed size, and formation process. All of the lagoons in this study are in a terraced coastal setting with high wave exposure; they vary predominantly in watershed size and the degree to which they include inherited space (that is, where the estuary is occupying a valley cut during times with lower sea levels). Based on this physical context, Jacobs et al. predict that San Dieguito Lagoon, the system with the largest watershed in our study, would be either perched above high tide or closed at high tide for approximately 70% of the time and open down to low in the intertidal for 10% of the time, with the remaining 20% spent in intermediate conditions (see table 1 in Jacobs et al. 2010). The other five lagoons were predicted to be either perched above high tide or closed at high tide 90% of the time. These predictions are broadly consistent with our results: in particular, that the lagoons were intermittently open systems, and that San Dieguito Lagoon was likely open more frequently than other lagoons. However, our findings, though inconclusive, suggest that the lagoons may have spent less time perched above high tide or closed at high tide than predicted by Jacobs et al. (e.g., based on textual descriptions of closure patterns and the assessment of historical sources showing open and closed conditions). These differences may be a result of generalizations made by the typology in Jacobs et al., which does not address site-specific variability among the five northern lagoons, and may also reflect biases in the historical record deriving from the timing (season and year) and source of observations.

Our findings are also consistent with findings from research on small estuaries and lagoons in other regions, which often have muted tides and are periodically closed by wave-built beach berms (Emmett et al. 2000). These systems are referred to by various terms, including ICOLLs (Intermittently Closed and Open Lake or Lagoon; Haines et al. 2006), temporarily open/closed estuaries, lagoons, blind estuaries (Teske and Wooldridge 2001), intermittent estuaries (Roy et al. 2001), periodically closed estuaries, and hypersaline estuaries (Day 1981). Systems vary in their closure frequency, ranging from predominantly open to predominantly closed, but their defining characteristic is that none are open all the time or closed all the time. These systems are found all over the world: they are the dominant estuarine type in some places, such as South Africa (Whitfield 2000) and New South Wales in southeastern Australia (Griffiths 1999) and are widespread in southern California (Ferren 1985, Jacobs et al. 2010, Dark et al. 2011). They are particularly prevalent in Mediterranean, semi-arid, or arid climates with highly seasonal freshwater inflow and relatively low rainfall.

The mechanisms for lagoon opening and closure involve complex interactions between coastal and watershed processes related to wave height and power, littoral sediment transport, tidal dynamics, and the timing and magnitude of watershed freshwater inputs. Certain drivers propel the lagoon toward closure: in particular, small tidal prism combined with low freshwater inflow and high wave energy tend to cause lagoon closure (Day 1981, Cooper 1990). In contrast, lagoon opening is largely influenced by watershed factors: lagoons open when freshwater runoff is high from storm events or floods, flooding the lagoon and raising water levels until the bar is breached (Day 1981, Cooper 1990, Roy et al. 2001, Elwany et al. 2003). Tidal flow, if sufficient, can also be responsible for maintaining an inlet for an extended period of time even in the absence of high freshwater flows.

Long-time resident Allan O. Kelly described this for Agua Hedionda Lagoon in the early 20th century:

> Following the 1927 floods, the channel remained open for more than five years and the sand bars and beaches along the mouth of the lagoon became very popular picnic places. Everybody wanted to swim in the water heated by the sun as it spread out over the hundreds of acres of mud flats east of the railroad (Kelly in Harmon 1967).

As a result of the importance of watershed dynamics in inlet formation in these systems, many intermittently closing systems have in common a highly seasonal flow regime and a relatively small watershed area; Cooper (1990) estimates that in southeastern Africa rivers with watersheds under about 500 km² (about 200 mi²) tend to support open conditions only during large seasonal flow events. This is consistent with trends observed in northern San Diego County, where all systems studied except San Dieguito Lagoon had watershed areas less than this threshold. Seasonal runoff patterns in North County would have tended to close lagoons during the summer and open them during wet-season storm events.

Nearshore bathymetry can also be an important factor influencing the frequency of opening (Bascom 1954). Breaching of the beach barrier following freshwater runoff is most likely to occur where there is a low spot on the barrier crest, which would be expected shoreward of submarine canyons where wave refraction would lead to a reduction in breaker height and wave uprush distance, and ultimately lowered berm height. Alternatively, a submarine ridge in the nearshore zone would lead to a convergence of wave orthogonals, an increase in breaker height, and a greater likelihood of breaching during storm wave attack (Wayne Engstrom, pers. comm.). The bathymetry immediately offshore of the study area lacks major submarine canyons or ridges, with the exception of the Carlsbad Canyon just southwest of Agua Hedionda Lagoon.

Over the past centuries, North County lagoons have experienced dramatic and distinct transformations in their closure dynamics as a result of modifications to the lagoons and their watersheds. Buena Vista Lagoon's closure patterns changed dramatically in 1940, when a weir was built permanently disconnecting the lagoon from the ocean. In contrast, Agua Hedionda Lagoon (as of 1954) and Batiquitos Lagoon (as of 1996) have been dredged and jettied to maintain a continuous tidal connection. Other lagoons have experienced more complex transformations, and their inlets are currently actively managed to maintain tidal exchange.

The construction of railroad berms in the early 1880s and subsequent road infrastructure undoubtedly had a significant impact on lagoon inlet dynamics by restricting inlet migration, changing scouring patterns from tides and floods, and altering sediment transport and deposition. Rather than precipitating a shift from fully tidal systems to closing systems, however, these modifications changed the frequency and timing of closure in already intermittently closing systems. In many cases, it appears that these constrictions produced more frequent closure conditions by the mid-20th century than previously observed (e.g., Mudie et al. 1976, Elwany et al. 1998, San Diego Regional Water Quality Control Board 1967). The restriction of channel and inlet migration by railroad and highway berms likely contributed to increased in-channel sediment accumulation, potentially decreasing tidal prism and runoff flow velocities and reducing the frequency of inlet opening (Webb et al. 1991). Other 20th century modifications, including flow regulation from dams, wastewater discharge, and increased sedimentation in the estuary from land use changes, contributed to changes in inlet dynamics in the 20th century and/or exacerbated issues associated with more frequent lagoon closure, such as increased flood risk and impaired water quality.

Conceptual Synthesis

The research presented in this report documents the historical ecological and hydrologic patterns of northern San Diego County lagoons. These systems exhibit characteristics similar to those observed in other estuaries with small watersheds and in climates with low and highly seasonal rainfall: most notably, dynamic conditions characterized by seasonal and annual variability in inlet condition and wide fluctuations in salinity from hypersaline to fresh/brackish (e.g., Cooper 1990, Largier 1997, Roy et al. 2001, Cooper 2001, Lichter et al. 2011).

Like all estuaries, North County lagoons are influenced by both coastal and watershed processes. Many of the environmental factors influencing lagoon character are relatively similar across systems, such as climate, tidal range, and wave exposure. However, other factors vary from lagoon to lagoon. In particular, the nature of fluvial flow and sediment varies from system to system, informed by each estuary's geophysical template (e.g., watershed size, topography, and geology). Though coastal processes clearly influence lagoon ecology and dynamics, understanding the watershed processes that shape them is just as – if not more – important for these system types (Elwany et al. 1998, Rich and Keller 2012).

This section aims to describe the primary components of the salt marsh and salt flat dominated systems historically found in northern San Diego County, and to integrate the ecological findings presented in this report with an understanding of key physical factors that influenced the systems: namely water and sediment, from both coastal/tidal and terrestrial/fluvial processes. Though a quantification of these processes is outside the scope of this project, the following discussion gives a coarse sense of how these processes varied across lagoon surfaces and over time. A second goal is to illustrate the extreme seasonal variability in lagoon character as they respond to changes in climate, hydrology, and sediment transport. Note that these illustrations are intended to be most representative of the salt flat-dominant systems, though many of the processes and landforms described are relevant for all six lagoons.

Estuarine Morphology

Salt flat/salt marsh-dominated lagoons were characterized by three primary features, each representing different formation processes and exhibiting different sediment and salinity characteristics. Moving inland from the ocean, these landforms are the marsh plain/flood-tide delta, the central salt flat, and the fluvial delta. Similar, though not identical, patterns are described in intermittently closing estuaries in South Africa (Cooper 2001, 2002) and southeastern Australia (Roy et al. 2001) as well as barrier estuaries in Washington State (Shipman 2008). Each of these features is described briefly below.

MARSH PLAIN/FLOOD-TIDE DELTA The western edge of each lagoon supported a broad marsh plain crossed by channels and mud flats. Though there are few data on sediment texture of the marsh plain, early descriptions suggest a relatively coarse substrate driven by coastal sediment brought in by wind, waves, and tides, at least at the westernmost edges of the marsh plain. Purer (1942) noted the soil along western edge of San Elijo Lagoon was "sandy, as it has been blown in from the beach," and that in Agua Hedionda Lagoon there was an "extensive deposit of gravel with large rounded stones on the surface of portions of the marsh." These observations are consistent with observed flood-tide delta forming processes bringing littoral sand into the estuary.

CENTRAL SALT FLAT A large, unvegetated salt flat was found in the central part of many North County lagoons, between the marsh plain and the upslope alluvial fan. The salt flat was seasonally flooded, with large fluctuations in salinity, inundation depth, and extent (see page 150 for more details). The salt flat was underlain by fine-grained soils, described as "black loam or A-do-be" (Rodgers 1887-8a) and "sticky" or "heavy" clay (Holmes and Pendleton 1918, Storie and Carpenter 1929a,b).

ALLUVIAL FAN At the upstream end of each estuary, a fan or delta built by fluvial processes supported transitional brackish and freshwater wetland habitats. These features were higher and coarser than the central salt flats, and were characterized by loamy soils (Storie and Carpenter 1929a,b, Purer 1942).

Seasonal and Interannual Variability

Lagoon conditions varied seasonally and interannually, tracking fluctuations in freshwater inflow, waves, and sediment delivery. On one side of the lagoon, tides inundated the lagoon when the inlet was breached or when waves overtopped the beach barrier, and coastal sediment built beach berms and flood-tide deltas. On the other side, freshwater flow from creeks and runoff fed into the lagoon, introducing sediment from the watershed. Water entered the lagoon through rainfall and seepage, and left through evaporation and filtration.

The interplay between these factors is complex and dynamic, and would have resulted in somewhat unpredictable patterns in the timing and duration of lagoon closure and flooding across different years and systems. While in some years a lagoon might breach in the winter and close in the summer, in other years it might stay open, or not open at all (e.g., Kelly 1959, Harmon 1967). Longtime Carlsbad resident Allan O. Kelly described some of this variability:

> In the early days, most of the lagoons along the coast were dry in the summer time. Sometimes because of dry seasons when the creeks failed to flow and at other times when over average rainfall filled the lagoons and the water broke out to the sea. It was only on the occasional year when the stream flow was just sufficient to fill the lagoon without running over, that there was much stagnant water left over at the end of the summer. (Kelly 1959)

Though the specific timing and duration of closure and flooding varied, overall patterns can be discerned. The annotated graphics on the following pages synthesize the lagoons' historical ecological patterns and illustrate the different "phases" that existed depending on the availability and movement of sediment and water. A graphic summarizing these dynamics can also be found in Chapter 1 (see pages 14-15).

DRY PHASE (LATE SUMMER & FALL)
inlet closed, low inflow, lagoon dries up

During the dry season, the lack of flows into the lagoon coupled with littoral sediment deposition onshore would tend to close or restrict the inlet. Once closed, evaporation rates greatly exceeding freshwater input would tend to dry out the lagoon, leading to hypersaline conditions and crystallizing salts.

CHANNELS AND PONDS

Contain water trapped after inlet closure

No surface connection to ocean

Subsurface flow through beach berm results in relatively high local water table, often allowing features to retain water through dry season

Relatively limited tidal channel network; some channels and ponds are relict features reflecting inactive channels

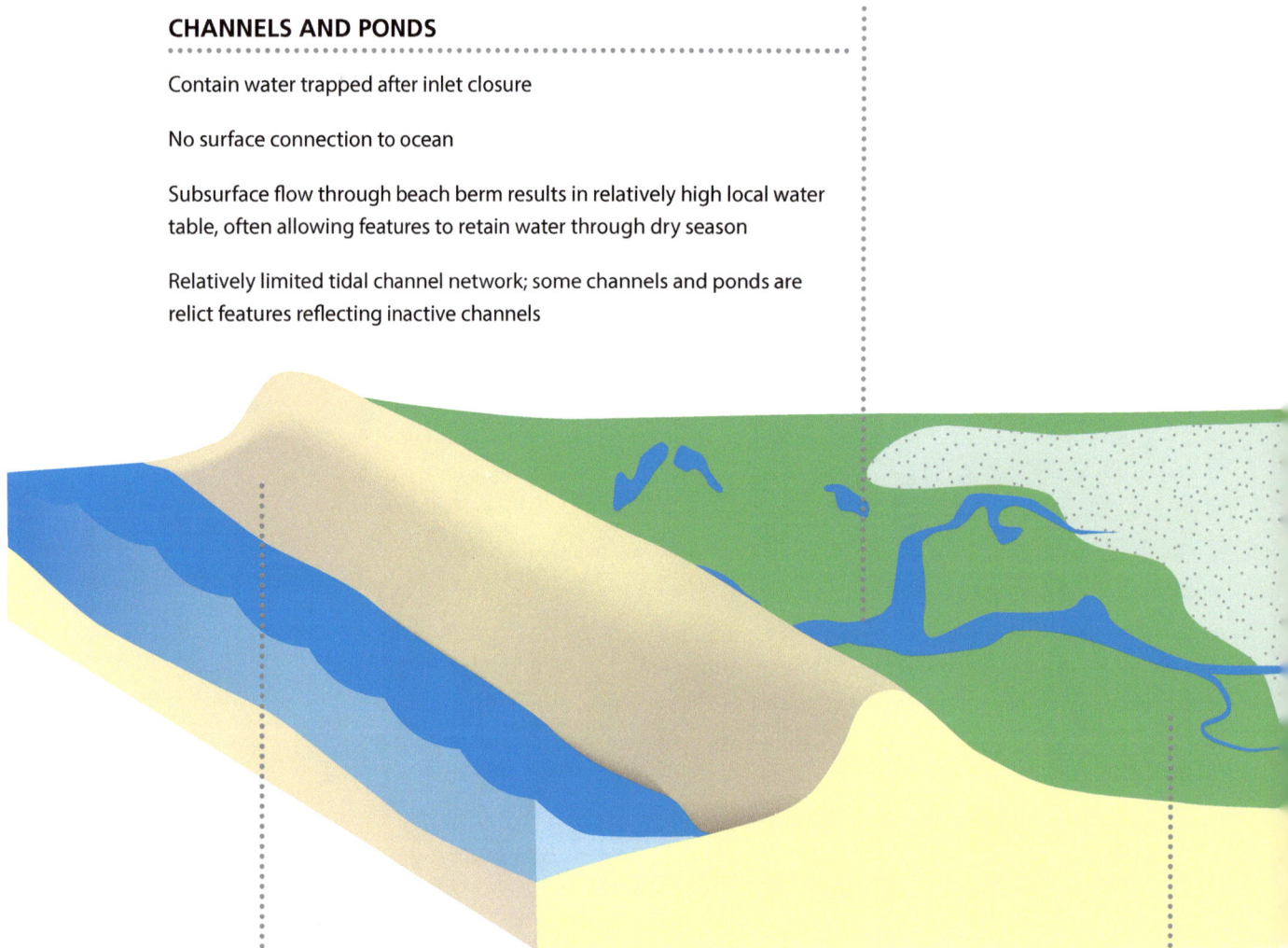

BEACH BERM

Inlet closes when sediment deposition from onshore wave action exceeds scouring due to flow through inlet

Berm is maintained by onshore wave action, blocking lagoon mouth and tidal exchange

Subsurface flow direction through berm is predominantly from ocean into lagoon when lagoon water level is below ocean level

SALT MARSH

Surface drained of tidal water following inlet closure and drying of lagoon

Relatively high local water table keeps marsh somewhat saturated through dry season

SEASONALLY FLOODED SALT FLAT

Dries out fully or partially, leaving unvegetated salt flat

Minimal to no inflow from precipitation and runoff; no connection to ocean

Evaporation greatly exceeds inflow, resulting in a desiccated surface and salt deposition

Hypersaline soil conditions preclude vegetation

FRESHWATER/BRACKISH WETLANDS

Subsurface flow from surrounding watershed results in relatively high local water table, even in summer

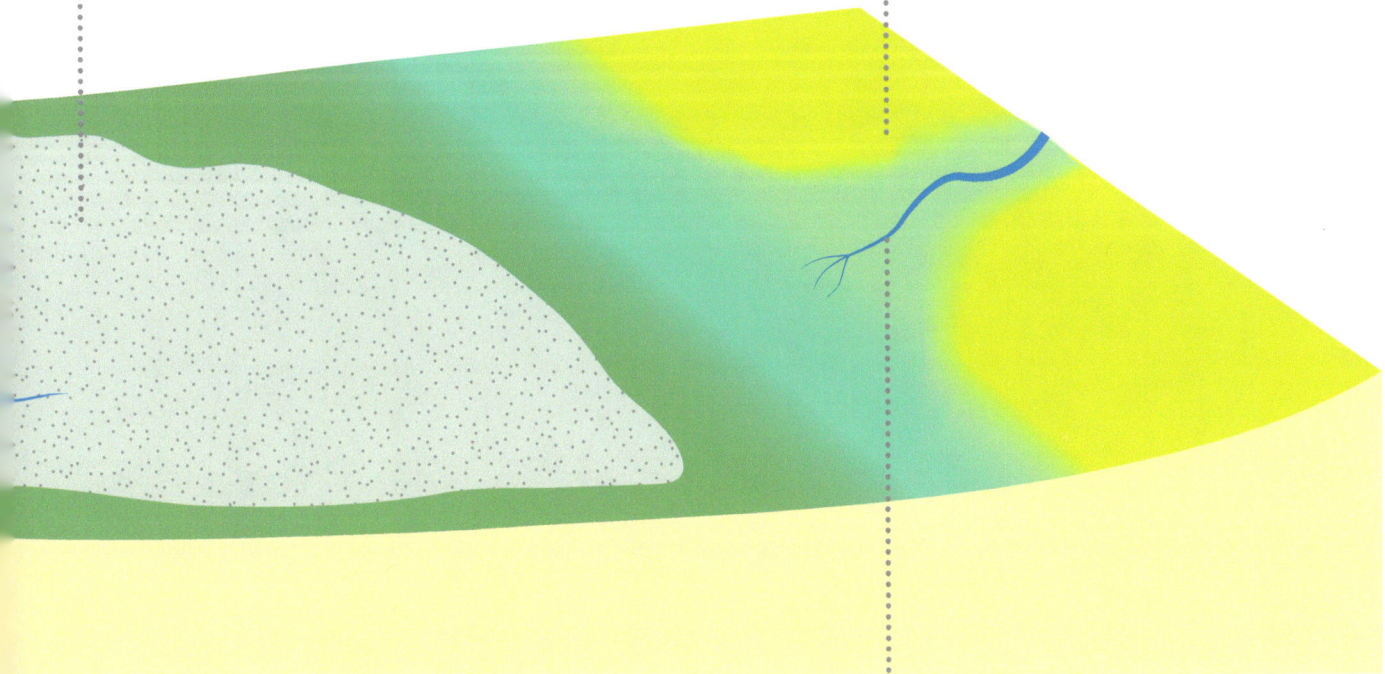

FLUVIAL CHANNEL

Little to no surface flow from the upstream watershed

No sediment transport to lagoon and upslope wetlands

DRY TO WET PHASE (EARLY WINTER)
inlet closed, stream flow fills lagoon

With the onset of rains, runoff and precipitation would begin to fill the lagoon with fresh water, impounding behind the beach berm and often creating perched conditions (that is, where the lagoon water level is above high tide). Instead of a hypersaline salt flat, a freshwater/brackish lagoon would form in the central portion of the estuary.

CHANNELS, PONDS, AND SALT MARSH

Stream inflow and precipitation raise water elevations, potentially flooding marsh

"Perched" conditions (lagoon levels above high tide)

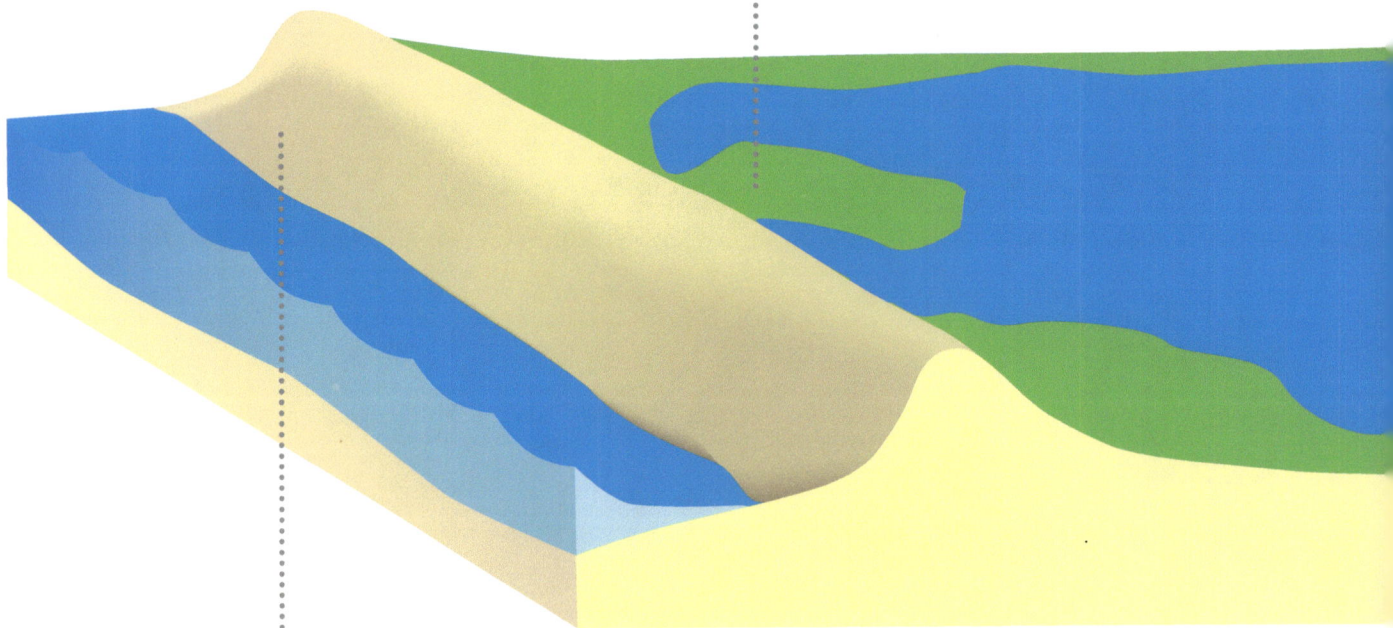

BEACH BERM

Berm maintained by onshore wave and aeolian transport; blocking lagoon mouth and restricting tidal exchange

Subsurface flow direction switches from ocean to lagoon (when lagoon levels low) to lagoon to ocean (when lagoon is perched)

SEASONALLY FLOODED SALT FLAT

Stream inflow and precipitation saturate surface, creating lagoon flooded with fresh to brackish water

Evaporation roughly balances inflow

FRESHWATER/BRACKISH WETLANDS

Stream flow saturates surface and raises local groundwater levels

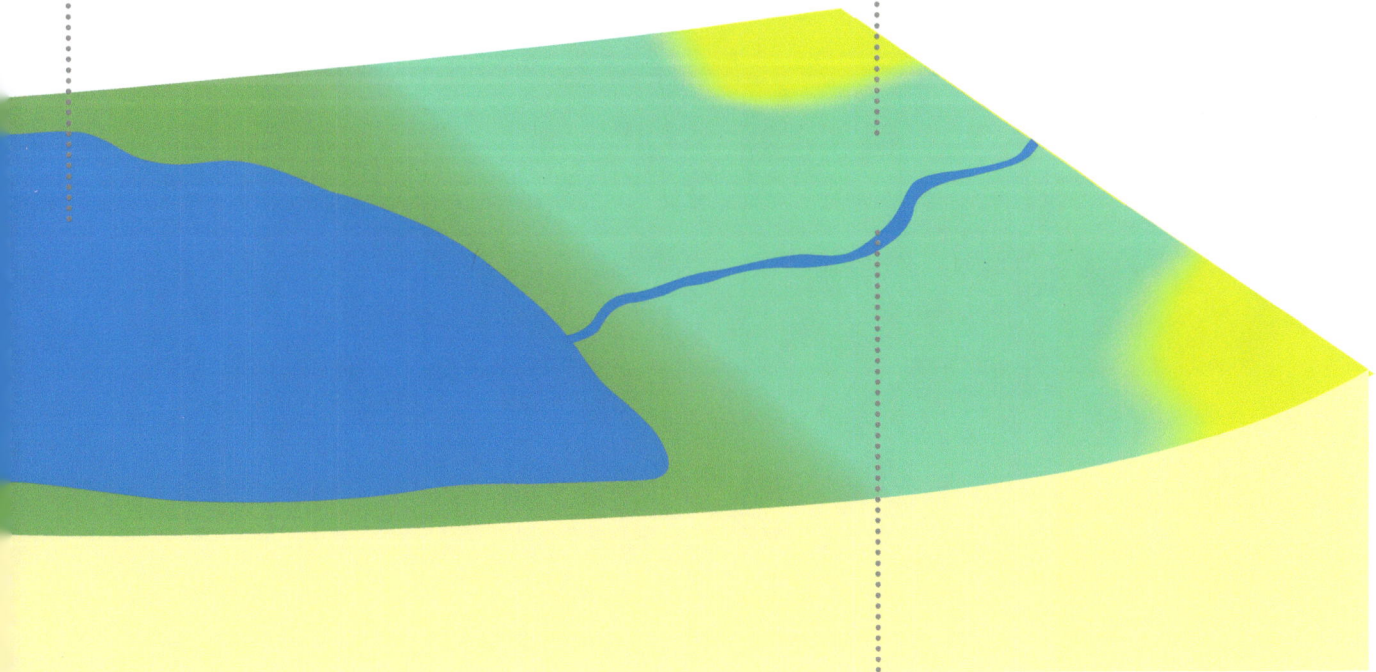

FLUVIAL CHANNEL

Stream flow initiated by early season rain

Surface flow reaches lagoon

WET PHASE (MID- TO LATE WINTER)
inlet opens, tidal conditions

Rainfall of sufficient magnitude would lead to inflow that could scour the lagoon and breach the beach barrier, forming an inlet and draining the lagoon. The lagoon could also be breached through a combination of elevated water levels in the estuary and overtopping by large waves. The lagoon would be subject to tidal exchange for a period of time, ranging in duration and depth of opening depending on the year and system.

SALT MARSH

Tidal sediment deposition builds flood-tide delta

Surface inundated when tide at or above MHHW

Relatively high local water table maintains saturated conditions at low tide

Soil salinity variable, but within range sufficient to sustain plant growth

BEACH BERM AND INLET

Inlet breaches when storm-induced high stream flow fills lagoon beyond capacity

Breached inlet allows stream flow out and daily tidal flow into and out of lagoon

Inlet may breach in different locations during different years and events

SEASONALLY FLOODED SALT FLAT

Modest tidal and fluvial sediment deposition and/or scouring during large storm events

Surface tidally inundated when tide at or above MHW

Also flooded by backup of stream inflow

FRESHWATER/BRACKISH WETLANDS

Relatively high local water table maintains saturated conditions

Fluvial sediment deposition during large storm events

Lower portions may be inundated by extreme high tides

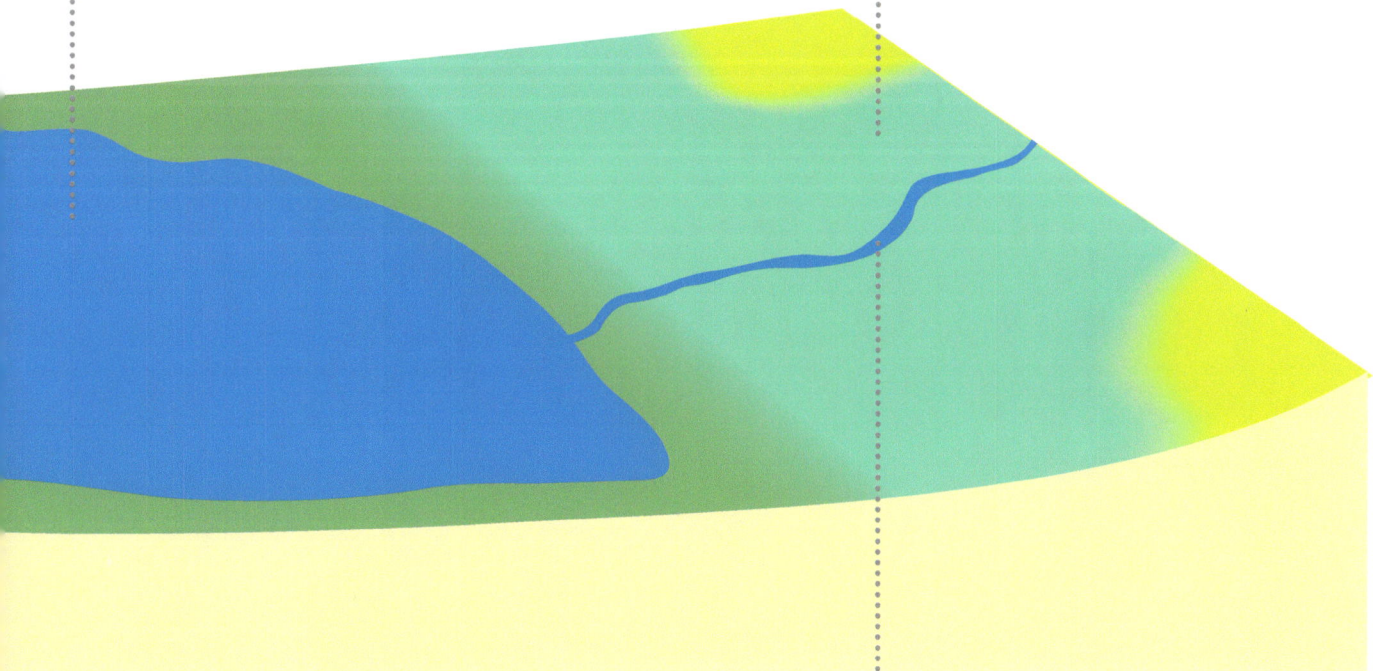

FLUVIAL CHANNEL

High baseflow with storm-induced high flows that deliver water and sediment

Stream overflows banks and spills onto floodplain

Surface flow reaches lagoon

WET TO DRY PHASE (SPRING & EARLY SUMMER)
inlet closes

As winter rains receded and inflow declined, wave action would again close the inlet, cutting off tidal exchange to the lagoon. Lagoon water levels could drop, remain steady, or even rise depending on net water balance (whether evaporation exceeded residual inflow). If water levels dropped, salinity would increase and salt flats would start to form.

CHANNELS AND PONDS

Contain water trapped after inlet closure

No surface connection to ocean

Seawater trapped at depth, with stratification and hypoxia

Subsurface flow through beach berm results in relatively high local water table, keeping features largely flooded through dry season

BEACH BERM

Inlet closes when sediment deposition from onshore wave action exceeds scouring from outflow through inlet, cutting off tidal exchange

SALT MARSH

Surface drained of tidal water when inlet closes

SEASONALLY FLOODED SALT FLAT

Lagoon water levels drop as water begins to evaporate from salt flat

Water salinities begin to increase

FRESHWATER/BRACKISH WETLANDS

Subsurface flow from surrounding watershed results in relatively high local water table, even during dry season

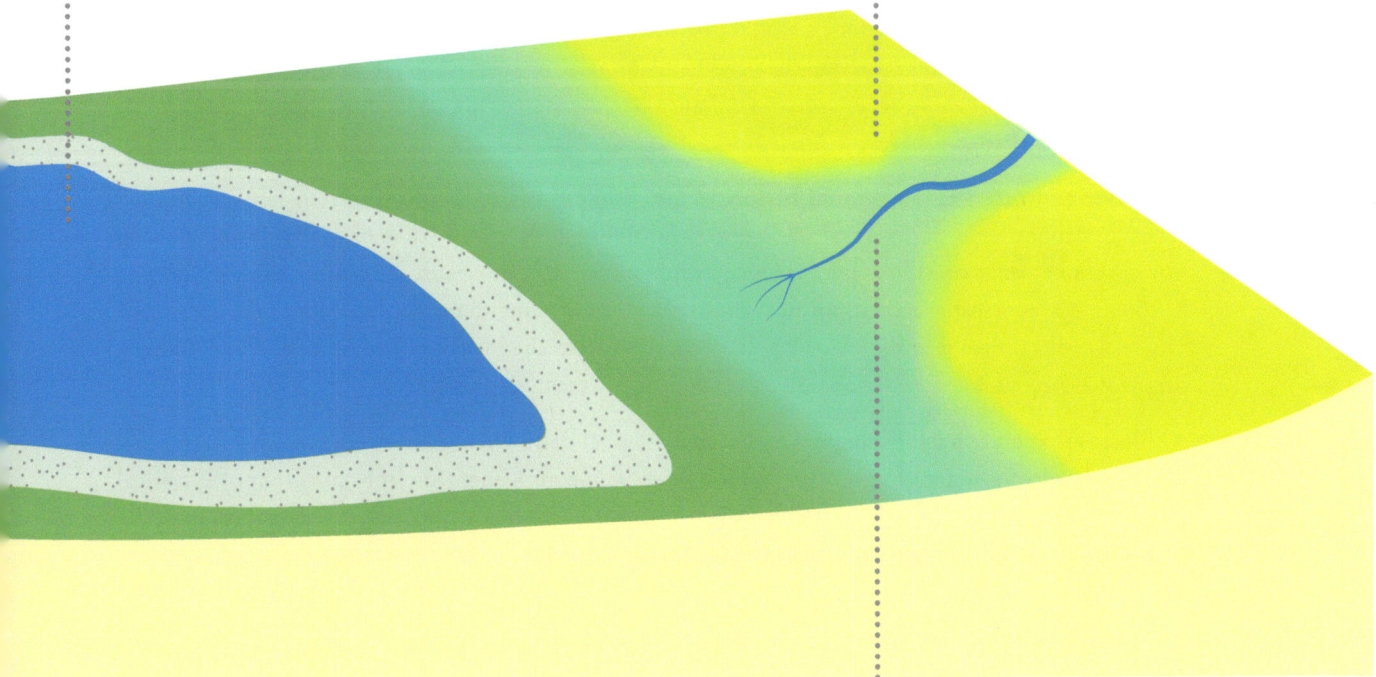

FLUVIAL CHANNEL

Decreasing surface flow and sediment transport from the upstream watershed

Recommended Future Research

Numerous avenues for further research would complement and expand on the work presented in this study. Development of these topics would enrich our understanding of the historical landscape and enhance our ability to apply these findings to current management activities and restoration planning by providing a more comprehensive understanding of system history, evolution, and response to physical drivers. A few recommendations for future research are described below.

- **Conduct historical ecology assessments for the watersheds draining into each lagoon.** While data relating to historical watershed characteristics and processes were assessed to some degree in this study, more detailed analysis would contribute to an understanding of fluvial/terrestrial inputs, groundwater dynamics, and the effects of land use changes on the lagoons.

- **Conduct historical ecology assessments for adjacent systems** (e.g., the San Luis Rey and Santa Margarita rivers and Loma Alta Slough). Understanding the historical character of these neighboring systems will provide insight into the patterns and variability of estuarine system types in northern San Diego County, enriching our understanding of the six systems described here.

- **Develop conceptual and/or numerical models** to link system evolution and observed historical and contemporary ecological patterns to hydrologic and geomorphic dynamics and changes over time (e.g., rising sea level, sediment dynamics, inlet closure and tidal circulation, coastal transgression, and anthropogenic impacts). These models would provide a more mechanistic understanding of the physical drivers shaping ecological patterns, shed light on the relative significance of anthropogenic changes to these drivers on habitat structure, and enhance our understanding of potential system response to future climate change.

- **Discover and interpret additional historical and ethnographic data** to better answer outstanding questions about regional historical ecological conditions and landscape dynamics. Promising avenues for additional data collection may include conducting interviews with long-time residents, sifting through local newspaper records for relevant material, continuing to pursue leads that may lead to additional early railroad records, reviewing diary accounts from the Mexican-American War, and examining USCGS records at the National Archives and/or Smithsonian Institution.

- **Collect and analyze paleoecological and archaeological data.** These data would provide information about longer-term system trajectories that would greatly complement the interpretation of historical archival data presented here. In particular, the strategic collection of additional core data from multiple lagoons and locations would provide a more robust understanding of sediment dynamics, sediment accretion rates, inlet closure dynamics, and vegetation community composition for northern San Diego County lagoons during both the prehistoric and historical period. Further analysis of existing core data taken for other purposes, such as Caltrans boring logs collected in the 1960s along the I-5 corridor prior to highway construction, could also yield useful insights. Core data could be used to create stratigraphic cross-sections of the lagoons depicting depth to bedrock, variations in sediment texture, and other subsurface patterns.

The research presented here provides the foundation to support local restoration and management efforts with robust historical data. However, this report alone is insufficient to make decisions about how to manage these systems. Report findings should be calibrated with assessments of contemporary conditions and with projected future climatic changes to develop practical, feasible, and spatially specific visions for future planning. Partnerships between local residents, managers, scientists, and other stakeholders is a crucial component of determining how to apply these data.

REFERENCES

Adams F. 1946. *The historical background of California agriculture*. Berkeley, CA: University of California Press.

Adams F, Harding ST, Robertson RD, et al. 1912. *Reports on the Irrigation Resources of California*. Sacramento, CA: FW Richardson, Superintendent of State Printing.

AECOM. 2012. Existing vegetation, San Elijo Lagoon restoration project. *Courtesy of San Elijo Lagoon Conservancy.*

Agnew DC, Legg M, Strand C. 1979. Earthquake history of San Diego. In *Earthquakes and Other Perils, San Diego Region*, 123-138: P.L. Abbott and W.J. Elliott, eds. San Diego, CA: San Diego Association of Geologists.

Ahnelt H, Göschl J, Dawson MN, et al. 2004. Geographical variation in the cephalic lateral line canals of *Eucyclogobius newberryi* (Teleostei, Gobiidae) and its comparison with molecular phylogeography. *Folia Zoologica* 53(4):385-398.

Akçakaya RH, Atwood JL, Breininger D, et al. 2003. Metapopulation dynamics of the California Least Tern. *The Journal of Wildlife Management* 67(4):829-842.

Al-Farraj A. 2005. An evolutionary model for sabkha development on the north coast of the UAE. *Journal of Arid Environments* 63:740-755.

Anderson D. 2007. Carlsbad: Rancho Agua Hedionda. http://www.carlsbad.ca.us/hedionda.html. Accessed August 27, 2014.

Anderson K. 2005. *Tending the wild: Native American knowledge and the management of California's natural resources*. Berkeley, CA: University of California Press.

Andrews E, Antweiler R, Neiman PJ, et al. 2004. Influence of ENSO on flood frequency along the California coast. *Journal of Climate* 17(2):337-348.

Applegate J. 1985. *Buena Vista Lagoon watershed sediment control study for the California Coastal Conservancy*. June Applegate & Associates Civil Engineers, Carlsbad, CA.

Aschmann H. 1966. *The natural & human history of Baja California*. Los Angeles: Dawson's Book Shop.

Bailey GE. 1902. *The saline deposits of California*. San Francisco, CA: California State Mining Bureau.

Baird M. 1904. *Oceanside, California, the gateway city of San Diego County*. Oceanside, CA: Oceanside Board of Trade.

Baker M. 1979. Morphological correlates of habitat selection in a community of shorebirds (Charadriiformes). *Oikos* 33:121-126.

Ball N. 1976. Lagoons noted. *The San Diego Union*. September 21. *Courtesy of San Diego Public Library.*

Bascom W. 1954. The control of stream outlets by wave refraction. *Journal of Geology* 62(6):600-605.

Battalio B, Danmeier D, Williams P. 2006. *Predicting closure and breaching frequencies of small tidal inlets - A quantified conceptual model*. 30th International Coastal Engineering Conference, San Diego, CA, USA.

Beasley T, Schuyler JD. 1889. Official map of San Diego County, California: compiled from latest official maps of U.S. surveys, railroad and irrigation surveys, county records and other reliable sources. [California]. *Courtesy of The Bancroft Library, UC Berkeley.*

Bell WA. 1869. *New tracks in North America*. S.I.: Chapman and Hall.

Beller EE, Grossinger RM, Salomon MN, et al. 2011. *Historical ecology of the lower Santa Clara River, Ventura River, and Oxnard Plain: an analysis of terrestrial, riverine, and coastal habitats. SFEI contribution #641*. San Francisco Estuary Institute, Oakland, CA.

Benton FW. 1886. *Zig-zag sketches of semi-tropic California and Las Peñasquitas.* Los Angeles, CA: Express Printing Co.

Bilotta GS, Brazier RE, Haygarth PM. 2007. The impacts of grazing animals on the quality of soils, vegetation, and surface waters in intensively managed grasslands. *Advances in Agronomy* 94:237-280.

Bliss NB. 2002. Cropland by county since 1850 and population since 1790 (cty2mc). U.S. Geological Survey (USGS) Land Cover Institute (LCI). http://landcover.usgs.gov/cropland/.

Bliss RS. 1846-7. *Copy of the diary of Robert S. Bliss, company B, Mormon Battalion, U.S. Army, 1846-1847.* [no publisher given] *Courtesy of the San Diego History Center.*

Bond MH. 2006. *Importance of estuarine rearing to central California steelhead* (Oncorhynchus mykiss) *growth and marine survival.* Master's thesis, Ecology and Evolutionary Biology, University of California, Santa Cruz. Santa Cruz, CA.

Boston KG. 1983. The development of salt pans on tidal marshes, with particular reference to south-eastern Australia. *Journal of Biogeography* 10(1):1-10.

Bowman JN. 1947. The area of the mission lands. *Courtesy of The Bancroft Library, UC Berkeley.*

Bradshaw JS. 1968. *Report on the biological and ecological relationship in the Los Peñasquitos Lagoon and Salt Marsh Area of the Torrey Pines State Reserve.* University of San Diego. *Courtesy of San Diego Public Library.*

Bradshaw JS, Mudie PJ. 1972. Some aspects of pollution in San Diego County lagoons. *California Marine Resource Commission, California Cooperative Oceanic Fisheries Investigations* 16:84-94.

Briere PR. 2000. Playa, playa lake, sabkha: Proposed definitions for old terms. *Journal of Arid Environments* 45:1-7.

Bronson JA. 1968. *Changing land use in the San Dieguito River Valley, San Diego County.* Master of Arts, Geography, San Diego State College.

Brownlie WR, Taylor BD. 1981. *Sediment management for southern California mountains, coastal plains and shoreline.* Part C. Coastal sediment delivery by major rivers in southern California. Pasadena, CA: California Institute of Technology.

Buena Vista Lagoon Foundation. n.d. State park/ecological reserve designation. http://buenavistalagoon.org/projects. html#statepark. Accessed January, 2014.

Burcham LT. 1956. *Historical geography of the range livestock industry of California.* Ph.D. dissertation, Geography, University of California, Berkeley, CA.

Burcham LT. 1961. Cattle and range forage in California: 1770-1880. *Agricultural History* 35(3):140-149.

Butler G. 1919. Map of San Diego County, California. *Courtesy of San Diego County Assessor.*

Byrd BF. n.d. The archaeology around San Elijo Lagoon. San Elijo Lagoon Conservancy. http://www.sanelijo.org/archaeology-report. Accessed January, 2014.

Byrd BF. 2004. The context and approach to the project. In *Results of NSF-funded archaeological and paleoenvironmental investigations at San Elijo Lagoon, San Diego County, California,* ed. Brian F. Byrd, Kevin O. Pope, and Seetha N. Reddy, 3-14. Carlsbad, CA.

Byrd BF, Reddy SN. 2002. Late Holocene adaptations along the northern San Diego coastline. In *Catalysts to complexity: Late Holocene of the California coast,* ed. J. Erlandson and T.L. Jones, 41-62: Cotsen Institute of Archaeology, University of California, Los Angeles.

Cain WC. 1982. *Geologic investigation and soil analysis of Buena Vista Lagoon, Oceanside, San Diego County.* California Department of Transportation.

California Department of Public Works. 1949. *San Dieguito and San Diego Rivers investigation.* Department of Public Works, Division of Water Resources Bulletin no. 55.

California Department of Transportation (Caltrans) and San Diego Association of Governments (SANDAG). 2013. *North coast corridor public works plan/transportation and resource enhancement program.*

California Development Board et al. 1923. *Agricultural and soil survey of San Diego County, California. Courtesy of California Society of Pioneers.* San Diego, CA: Press of Frye & Smith.

California State Data Center. 2012. Historical census populations of counties and incorporated cities in California, 1850-2010. http://www.dof.ca.gov/research/demographic/state_census_data_center/historical_census_1850-2010/.

Callaway RM, Jones S, Wayne FRJ, et al. 1990. Ecology of a mediterranean-climate estuarine wetland at Carpinteria, California: plant distributions and soil salinity in the upper marsh. *Canadian Journal of Botany* 68:1139-1146.

Capace N. 1999. *Encyclopedia of California.* St. Clair Shores, MI: Somerset Publishers Inc.

Capelli MH. 1997. *Tidewater goby (Eucyclogobius newberryi) management in California estuaries.* Proceedings, California and the World Ocean Conference. San Diego, California.

Carlsbad Watershed Network. 2002. *The Carlsbad Watershed Management Plan.* http://www.projectcleanwater.org/html/ws_carlsbad_plan_network_plan.html

Carpelan LH. 1969. Physical characteristics of southern California coastal lagoons. In *Lagunas costeras, un simposio*, ed. A. Ayala Castanares and F. B. Phleger, 319-334. Mexico, D.F.: Universidad Nacional Autónoma de México.

Carpenter, FA. 1913. *The climate and weather of San Diego, California.* San Diego, CA: San Diego Chamber of Commerce.

Carpenter NK. 1939. ORNIS, record for *Sterna albifrons browni* from "Del Mar, At mouth of San Dieguito River." Western Foundation of Vertebrate Zoology (WFVZ).

Carrico RL. 1977. Portola's 1769 expedition and coastal native villages of San Diego County. *Journal of California Anthropology* 4(1):30-41.

Chase JS. 1913. *California coast trails: a horseback ride from Mexico to Oregon.* Boston, MA and New York, NY: Houghton Mifflin Company.

Christenson LN, Sweet EL. 2008. *Ranchos of San Diego County.* Charleston, SC: Arcadia Publishing.

Cleland RG. [1941]1990. *The cattle on a thousand hills: Southern California, 1850-1880.* San Marino, CA: The Huntington Library.

Coats RN, Swanson M, Williams P. 1989. Hydrologic analysis for coastal wetland restoration. *Environmental Management* 13(6):715-727.

Coats RN, Williams PB, Cuffe CK, et al. 1995. *Design guidelines for tidal channels in coastal wetlands.* San Francisco, CA. Available at http://www.esassoc.com/sites/default/files/DesignGuidelns-TidalWetlnds-RevSept24_02.pdf.

Cole KL, Wahl E. 2000. A late Holocene paleoecological record from Torrey Pines State Reserve, California. *Quaternary Research* 53:341-351.

Cooke PSG. 1849. Report from the Secretary of War, communicating a copy of the official journal of Lieutenant Colonel Philip St. George Cooke, from Santa Fé to San Diego, &c. In *Public documents printed by order of the Senate of the United States, during a special session begun and held at the City of Washington, March 5, 1849*, ed. Washington, D.C.: Union Office.

Cooper JAG. 1990. Ephemeral stream-mouth bars at flood-breach river mouths on a wave-dominated coast: Comparison with ebb-tidal deltas at barrier inlets. *Marine Geology* 95:57-70.

Cooper JAG. 2001. Geomorphological variability among microtidal estuaries from the wave-dominated South African coast. *Geomorphology* 40:99-122.

Cooper JAG. 2002. The role of extreme floods in estuary-coastal behaviour: contrasts between river- and tide-dominated microtidal estuaries. *Sedimentary geology* 150:123-137.

Costansó M, Browning P. 1992. *The discovery of San Francisco Bay: the Portola expedition of 1769-1770: the diary of Miguel Costanso, in Spanish and English = El descubrimiento de la Bahia de San Francisco: la expedicion de Portola de 1769-1770.* Edited by Peter Browning. Lafayette, CA: Great West Books.

County of San Diego. 1970. *The coastal lagoons of San Diego County.* Prepared by The Environmental Task Force, County of San Diego. *Courtesy of San Elijo Lagoon Conservancy.*

County of San Diego. 1974. *Draft environmental impact report: San Elijo Lagoon -- acquisition.* San Diego County Public Works Agency, Park Development Division.

County of San Diego. 1979. *Water and marine resources, environmentally sensitive habitat, and hazards. Courtesy of San Elijo Lagoon Conservancy.*

County of San Diego. 1996. *San Elijo Lagoon Area enhancement plan.* http://scc.ca.gov/webmaster/ftp/pdf/sccbb/2008/0812/0812Board12_San_Elijo_Lagoon_Ex3.pdf. Accessed August 4, 2014.

County of San Diego Department of Public Works. 2003. *San Diego County hydrology manual.* http://www.sdcounty.ca.gov/dpw/floodcontrol/floodcontrolpdf/hydro-hydrologymanual.pdf. Accessed August 4, 2014.

Couts CJ. 1887. Map of South Oceanside, San Diego Co. *Courtesy of San Diego County Assessor.*

Cowardin LM, Carter V, Golet FC, et al. 1979. *Classification of wetlands and deepwater habitats of the United States.* Fish and Wildlife Service, Biological Services Program, U.S. Department of the Interior. Washington, D.C.

Crabtree R, Warren C, True DL. 1963. *Archaeological investigations at Batiquitos Lagoon, San Diego County, California.* Archaeological Survey Annual Report:321-345. Department of Anthropology-Sociology, University of California, Los Angeles.

Crespí J, Bolton HE. 1927. *Fray Juan Crespí, missionary explorer on the Pacific coast, 1769-1774*. Berkeley, CA: University of California Press.

Crespí J, Brown AK. 2001. *A description of distant roads: original journals of the first expedition into California, 1769-1770*. San Diego, CA: San Diego State University Press.

Crimmins D, 1990. Lagoon source of beauty, controversy. *North County Blade-Citizen*. April 15. *Courtesy of Encinitas Historical Society*.

Crooks J, McCullough J, Bellringer H, et al. 2012. *The physical, chemical and biological monitoring of Los Penasquitos Lagoon*. Southwest Wetlands Interpretive Association.

CSUN (California State University, Northridge) Center for Geographical Studies. 2012. [Wetland and riparian habitat mapping for southern California coastal watersheds.] Southern California Wetlands Mapping Project. http://www.socalwetlands.com/website/main.htm. Accessed April 16, 2013.

Dailey MD, Hill B, Lansing N. 1974. *A summary of knowledge of the southern California coastal zone and offshore areas*. Southern California Ocean Studies Consortium. Long Beach, CA.

Daily Alta California. 1864. San Diego County and its resources. April 25. *Courtesy of California Digital Newspaper Collection*.

Dark S, Stein ED, Bram D, et al. 2011. *Historical ecology of the Ballona Creek watershed. Technical Report #671*. Southern California Coastal Water Research Project, Costa Mesa, CA.

Dawson MN, Staton JL, Jacobs DK. 2001. Phylogeography of the tidewater goby, *Eucyclogobius newberryi* (teleostei, gobiidae), in coastal California. *Evolution* 55(6):1167-1179.

Day JH. 1981. *Estuarine ecology with particular reference to southern Africa*. Rotterdam: A.A. Balkema.

de Anza JB, Bolton HE. 1930. *Anza's California expeditions*. Berkeley, CA: University of California Press.

de Mofras D, Wilbur ME. 1937. *Duflot de Mofras' travels on the Pacific Coast, Vol. II*. Santa Ana, CA: The Fine Arts Press.

del Barco M, Tiscareno F, Leon-Portilla M, et al. [ca. 1770]1980. *The natural history of Baja California*. Los Angeles, CA: Dawson's Book Shop.

Desmond JS, Williams GD, Vivian-Smith G, et al. 2001. The diversity of habitats in southern California coastal wetlands. In *Handbook for restoring tidal wetlands*, ed. Zedler JB. Boca Raton, LA: CRC Press.

Dodge R. 1958. The Fallbrook Line. *Dispatcher* (17). http://www.sdrm.org/history/cs/calsouth.html

Duhaut-Cilly A. [1827]1997. *A voyage to California, the Sandwich Islands, and around the world in the years 1826-1829*. August Frugé and Neal Harlow. Berkeley, CA: University of California Press.

Earl DA, Louie KD, Bardeleben C, et al. 2010. Rangewide microsatellite phylogeography of the endangered tidewater goby, Eucyclogobius newberryi (Teleostei: Gobiidae), a genetically subdivided coastal fish with limited marine dispersal. *Consevation Genetics* 11:103-114.

Ellis AJ, Lee CH. 1919. *Geology and ground waters of the western part of San Diego County, California*. Water-supply paper (Washington, D.C.), no. 446. Washington, DC: Government Printing Office.

Elwany MHS, Thum AB, Gayman W, et al. 1995. *Historical evaluation of San Dieguito Lagoon and inlet*. SIO Reference, 95-19. La Jolla, CA: Center for Coastal Studies, Scripps Institution of Oceanography, University of California, San Diego.

Elwany MHS. 2011. Characteristics, restoration, and enhancement of southern California Lagoons. *Journal of Coastal Research* (59):246-255.

Elwany MHS, Flick RE, et al. 1998. Opening and closure of a marginal southern California lagoon inlet. *Estuaries* 21(2):246-254.

Elwany MHS, Flick RE, Hamilton MM. 2003. Effect of a small southern California lagoon entrance on adjacent beaches. *Estuaries* 26(3):700-708.

Emmett R, Llansó R, Newton J, et al. 2000. Geographic signatures of North American West Coast estuaries. *Estuaries* 23(6):765-792.

Engelhardt Z. 1920. *San Diego Mission*. San Francisco, CA: James H. Barry, Co.

Engelhardt Z. 1921. *San Luis Rey Mission*. San Francisco, CA: James H. Barry, Co.

Engstrom WN. 2006. Nineteenth century coastal geomorphology of Southern California. *Journal of Coastal Research* 22(4):847-861.

Everest International Consultants, Inc. 2004. *Buena Vista Lagoon restoration feasibility analysis: Final report*. Long Beach, CA.

Ewing NH. 1988. *Del Mar: looking back*. Del Mar, CA: Del Mar History Foundation.

Fages P, Priestley HI. 1937. *A historical, political, and natural description of California.* Berkeley, CA: University of California Press.

Ferren WR. 1985. *Carpinteria Salt Marsh: Environment, history, and botanical resources of a southern California estuary.* The Herbarium, Department of Biological Sciences, University of California, Santa Barbara.

Fetzer L. 2005. *San Diego County place names A to Z.* San Diego, CA: Sunbelt Publications, Inc.

Flanigan SK, Carrico S, Carrico R. 1993. *Oceanside, California's pride. 1992 cultural resource survey.* Oceanside, CA.

Flick RE, Murray JF, Ewing LC. 1999. *Trends in U.S. tidal datum statistics and tide range: A data report atlas.* Scripps Institution of Oceanography, La Jolla, California.

Flick R. 2005. Dana Point to the international border. In *Living with the changing California coast*, ed. Gary Griggs, Kiki Patsch, and Lauret Savoy. Berkeley, CA: University of California Press.

Foster J. 1853. No. 348, William Carey Jones. Ex Mission of San Luis Rey, map. Annexed to deposition of Juan Foster, December 30th, 1853, Geo. Fisher, scy. Land Case Map D-1386A. U.S. District Court, Southern District. *Courtesy of The Bancroft Library, UC Berkeley.*

Freeman JE. 1854a. *Field notes of the base lines in Townships 9, 10, 11, and 12 South, Ranges 4, 5, and 6 West, San Bernardino Meridian.* U.S. Department of the Interior, Bureau of Land Management Rectangular Survey, California. Book 170-15. *Courtesy of Bureau of Land Management, Sacramento, CA.*

Freeman JE. 1854b. *Field notes of the surveys of lines Townships 14, 15, and 16 South, Ranges 1, 2, 3, and 4 West, San Bernardino Meridian.* U.S. Department of the Interior, Bureau of Land Management Rectangular Survey, California. Book 167-31. *Courtesy of Bureau of Land Management, Sacramento, CA.*

Freeman WB, La Rue EC, Padgett HD. 1912. Part IX, Colorado River basin. In *Surface water supply of the United States, 1910*, ed., 1-233. Washington: Government Printing Office.

G. Bancroft Collection. 1932. ORNIS, record for *Charadrius alexandrinus nivosus* from "Del Mar, A mile north of Del Mar." Western Foundation of Vertebrate Zoology (WFVZ).

Gallegos D. 1992. Patterns and implications of coastal settlement in San Diego County: 9000 to 1300 years ago. In *Essays on the prehistory of maritime California*, ed. Terry Jones, 205-216. Davis, CA: Center for Archaeological Research at Davis.

Gallegos D. 2002. Southern California in transition: Late Holocene occupation of southern San Diego County, California. In *Catalysts to complexity: Late Holocene of the California coast*, ed. J. Erlandson and T.L. Jones. Los Angeles, CA: Cotsen Institute of Archaeology, University of California, Los Angeles.

Gander FF. 1936. Consortium of California Herbaria, record for *Suaeda taxifolia* from "Bluff at north side of mouth of San Dieguito Creek." San Diego Natural History Museum Herbarium.

Gayman W. 1978a. *Estimation of present and past tidal prisms in Batiquitos Lagoon.* Douglas L. Inman Papers, 1940-2007, Accession no. 2006-52, Box 13, USA-C2.3. *Courtesy of Scripps Institution of Oceanography Archives, UC San Diego.*

Gayman W. 1978b. *Preliminary report on historical and geological data bearing on the tidal character of Batiquitos Lagoon during the last two hundred years.* Douglas L. Inman Papers, 1940-2007, Accession no. 2006-52, Box 13, USA-C2.3. *Courtesy of Scripps Institution of Oceanography Archives, UC San Diego.*

Goldsworthy J. 1874a. *Field notes of the interior and part of exterior lines of Township 14 South, Range 4 West, San Bernardino Meridian.* U.S. Department of the Interior, Bureau of Land Management Rectangular Survey, California. Book 300-35. *Courtesy of Bureau of Land Management, Sacramento, CA.*

Goldsworthy J. 1874b. *Field notes of the part of the exterior and the interior lines of Township 14 South, Range 3 West, San Bernardino Meridian.* U.S. Department of the Interior, Bureau of Land Management Rectangular Survey, California. Book 169-23. *Courtesy of Bureau of Land Management, Sacramento, CA.*

Goodwin P, Fishbain L, Naugles G. 1992. *San Elijo Lagoon enhancement plan, second interim report.* Report to California Coastal Conservancy and San Diego Department of Parks and Recreation.

Greer K, Stow D. 2003. Vegetation type conversion in Los Peñasquitos Lagoon, California: An examination of the role of watershed urbanization. *Research* 31(4):489-503.

Grewell BJ, Callaway JC, Ferren Jr. WR. 2007. Estuarine wetlands. In *Terrestrial vegetation of California*, ed. Michael G. Barbour, Todd Keeler-Wolf, and Allan A. Schoenherr, 124-154. Berkeley, CA: University of California Press.

Griffiths SP. 1999. Consequences of artifically opening coastal lagoons on their fish assemblages. *International Journal of Salt Lake Research* 8:307-327.

Grossinger RM. 1995. *Historical evidence of freshwater effects on the plan form of tidal marshlands in the Golden Gate Estuary.* Master's thesis, Marine Sciences, University of California, Santa Cruz. Santa Cruz, CA.

Grossinger RM. 2005. Documenting local landscape change: the San Francisco Bay area historical ecology project. In *The historical ecology handbook: a restorationist's guide to reference ecosystems*, ed. Dave Egan and Evelyn A. Howell, 425-442. Washington, D.C.: Island Press.

Grossinger RM, Askevold RA. 2005. *Historical analysis of California Coastal landscapes: Methods for the reliable acquisition, interpretation, and synthesis of archival data. Report to the U.S. Fish and Wildlife Service San Francisco Bay Program, the Santa Clara University Environmental Studies Institute, and the Southern California Coastal Water Research Project, SFEI Contribution 396.* San Francisco Estuary Institute, Oakland, CA.

Grossinger RM, Stein ED, Cayce K, et al. 2011. *Historical wetlands of the southern California coast: an atlas of U.S. Coast Survey t-sheets, 1851-1889. SFEI contribution #586, SCCWRP technical report #589.* San Francisco Estuary Institute, Oakland, CA.

Grossinger RM, Striplen CJ, Askevold RA, et al. 2007. Historical landscape ecology of an urbanized California valley: wetlands and woodlands in the Santa Clara Valley. *Landscape Ecology* 22:103-120.

Gudde EG, Bright W. 1998. *California place names: the origin and etymology of current geographical names.* 4th ed., rev. and enl. edition. Berkeley, CA: University of California Press.

Gunn D. 1886. *San Diego. Climate, productions, resources, topography, etc., etc.* San Diego, CA: Union Steam Book and Job Print Office.

Gunn D. 1887. *Picturesque San Diego with historical and descriptive notes.* Chicago, IL: Knight and Leonard Co., Printers.

Gutierrez SS. 2002. *Windows on the Past: An Illustrated History of Carlsbad, California.* Virginia Beach, VA: Donning Co.

Haines PE, Tomlinson RB, Thom BG. 2006. Morphometric assessment of intermittently open/closed coastal lagoons in New South Wales, Australia. *Estuarine, Coastal and Shelf Science* 67:321-332.

Hall WH. 1888. *Irrigation in California (Southern): the field, water-supply and works, organization and operation in San Diego, San Bernardino and Los Angeles counties* Sacramento, CA: State Office, J.D. Young, Supt. State Printing.

Handford CR. 1981. A process-sedimentary framework for characterizing recent and ancient sabkhas. *Sedimentary Geology* 30:255-265.

Hanley N, Ready R, Colombo S, et al. 2009. The impacts of knowledge of the past on preferences for future landscape change. *Journal of Environmental Management* 90:1404-1412.

Hanson AP. 1880. *Field notes of the subdivision lines of Township 13 South Range 4 West, San Bernardino Meridian, California.* U.S. Department of the Interior, Bureau of Land Management Rectangular Survey, California. Book 170-18. *Courtesy of Bureau of Land Management, Sacramento, California.*

Harley JB. 1989. Historical geography and cartographic illusion. *Journal of Historical Cartography* 15:80-91.

Harmon J. 1967. *History of Carlsbad.* Carlsbad, CA: Friends of the Library.

Harrington JP. 1925. San Alejo estero. OPPS Neg 91-34493. John P. Harrington Papers. Photographs. Southern California/Basin, Diegueno. *Courtesy of National Anthropological Archives, Smithsonian Institution.*

Harrington JP. ca. 1930a. *Luiseño texts.* The Papers of John Peabody Harrington in the Smithsonian Institution, 1907-1957. Volume 3. Native American history, language, and culture of southern California/Basin. Ed. by Elaine L. Mills and Ann J. Brickfield. Kraus International Publications: Millwood, NY. 1986. Reel 121:27.

Harrington JP. ca. 1930b. *Luiseño vocabulary.* The Papers of John Peabody Harrington in the Smithsonian Institution, 1907-1957. Volume 3. Native American history, language, and culture of southern California/Basin. Ed. by Elaine L. Mills and Ann J. Brickfield. Kraus International Publications: Millwood, NY. 1986. Reel 117:560.

Harris GW, Simmons BJ. 1916. The A.T. and S.F. RY. Co. coast lines preliminary and location surveys up San Dieguito Valley near Del Mar. Carl Leavitt Hubbs (1894-1979) Papers, 1920-1979, Box 270, Folder 39. *Courtesy of Scripps Institution of Oceanography Archives, UC San Diego.*

Harris JA, Hobbs RJ, Higgs E, et al. 2006. Ecological restoration and global climate change. *Restoration Ecology* 14:170-176.

Harrison EN. 1932a. ORNIS, record for *Charadrius alexandrinus nivosus* from "Del Mar, On strand near beach." Western Foundation of Vertebrate Zoology (WFVZ).

Harrison EN. 1932b. ORNIS, record for *Sterna albifrons browni* from "Encinitas, First lagoon north of Encinitas." Western Foundation of Vertebrate Zoology (WFVZ).

Harrison EN. 1934a. ORNIS, record for *Charadrius alexandrinus nivosus* from "Encinitas, Three mi. north of Encinitas." Western Foundation of Vertebrate Zoology (WFVZ).

Harrison EN. 1934b. ORNIS, record for *Sterna albifrons browni* from "Encinitas, 3 mi. north of Encinitas." Western Foundation of Vertebrate Zoology (WFVZ).

Harrison EN. 1945. ORNIS, record for *Sterna albifrons browni* from "Encinitas, 2 mi. north of Encinitas." Western Foundation of Vertebrate Zoology (WFVZ).

Haskett W. 1999. Family grew up on edge of Batiquitos lagoon. *The San Diego Union-Tribune*. April 11. *Courtesy of Carlsbad City Library Carlsbad History Room.*

Hawthorne KS. 2003. *Highway 101. Location of the coast route between Oceanside and La Jolla*. Highway 101 Association. *Courtesy of Carlsbad City Library Carlsbad History Room.*

Hayes B. 1929. Pioneer notes from the diaries of Judge Benjamin Hayes, 1849-1875. Los Angeles, CA: priv. print. *Courtesy of Library of Congress.*

Hayes SA, Bond MH, Hanson CV, et al. 2008. Steelhead growth in a small central California watershed: upstream and estuarine rearing patterns. *Transactions of the American Fisheries Society* 137:114-128.

Hays JC. 1858a. *Field notes of the final survey of the Rancho Agua Hedionda, Juan Maria Marron, Confirmee*. Book 41. *Courtesy of Bureau of Land Management, Sacramento, CA.*

Hays JC. 1858b. *Field notes of the final survey of the Rancho Pueblo or Town Lands of San Diego, President and Trustees of the City of San Diego, Confirmees, by John C. Hays, Deputy Surveyor, under his instructions of June 30, 1858*. Book 35. *Courtesy of Bureau of Land Management, Sacramento, CA.*

Heaton L. 1923. ORNIS, record for *Charadrius alexandrinus nivosus* from "Cardiff, Near Cardiff." Western Foundation of Vertebrate Zoology (WFVZ).

Heilbron CH, editor. 1936. *History of San Diego County*. San Diego, CA: San Diego Press Club.

Higgs E. 2012. History, novelty, and virtue in ecological restoration. In *Ethical adaptation to climate change: human virtues of the future*, ed. Allen Thompson and Jeremy Bendik-Keymer, 81-102. Cambridge, MA: MIT Press.

Hinde HP. 1954. The vertical distribution of salt marsh phanerogams in relation to tide levels. *Ecological Monographs* 24(2):209-225.

Hinman W. 2012. For the birds: a brief history of Buena Vista Lagoon. *Carlsbad Magazine* 8(1):37-39.

Hoffman Bros. 1869. Map of San Diego County compiled from official surveys. Traced by GD Meek, January 28, 1937. San Francisco, CA. *Courtesy of San Diego History Center.*

Hoffman CF. 1869. *Field notes of the final survey of the Rancho San Diegito, Juliana L. Osuna, et al, Confirmee*. Book G-5. *Courtesy of Bureau of Land Management, Sacramento, CA.*

Holder CF. 1906. *Life in the open: sport with rod, gun, horse, and hound in southern California*. New York, NY: The Knickerbocker Press.

Holmes LC, Pendleton RL. 1918. *Reconnoissance soil survey of the San Diego region, California*. U.S. Department of Agriculture, Washington, D.C.

Holser WT, Javor B, Pierre C, et al. 1981. Geochemistry and ecology of salt pans at Guerrero Negro, Baja California. In *Geology of northwestern Mexico and southern Arizona*, ed. L. Ortlieb and J. Roldán, 1-56. Hermosillo, Sonora, Mexico: Estación Regional de Noroeste, Instituto de Geología, U.N.A.M.

Howard VE. 1869. *Pueblo lands of San Diego. Exceptions to survey made by John C. Hays, July 1858*. Filed in office of United States Surveyor General for California, May 19, 1869. San Francisco, CA: Mullin, Mahon & Co. *Courtesy of The Bancroft Library, UC Berkeley.*

Howard-Jones M. 1982. *Seekers of the Spring: A History of Carlsbad*. Carlsbad, CA: Friends of the Carlsbad Library.

Howard-Jones M. 1984. Between Germany and Carlsbad: A high yielding bond. *The Journal of San Diego History*. San Diego Historical Society Quarterly 30(2). http://www.sandiegohistory.org/journal/84spring/carlsbad.htm.

Hoyt F. 1954. San Diego's First Railroad: The California Southern. *Pacific Historical Review* 23(2):133-146.

Hubbs CL. 1921. *Latitudinal Variation in the Number of Vertical Fin-Rays in Leptocottus Armatus*. University of Michigan, Museum of Zoology.

Hubbs CL, Whitaker TW, Reid FMH. 2008. *Los Peñasquitos Marsh*. The Torrey Pines Association. http://www.torreypine.org/parks/penasquitos-lagoon.html.

Hughes C. 1975. The decline of the Californios: the Case of San Diego, 1846-1856. *The Journal of San Diego History* 21(3).

Inman D. 1983. Application of coastal dynamics to the reconstruction of paleocoastlines in the vicinity of La Jolla, California. In *Quaternary coastlines and marine archaeology*, ed. P. Masters and N. Flemming, 1-49. New York, NY: Academic Press.

Inman DL, Masters PM. 1991. *Budget of sediment and prediction of the future state of the coast*. U.S. Army Corps of Engineers, Los Angeles District.

Isola CR, Colwell MA, Taft OW, et al. 2000. Interspecific differences in habitat use of shorebirds and waterfowl foraging in managed wetlands of California's San Joaquin Valley. *Waterbirds* 23:196-203.

Iversen DR, Becker MS, Wolf JS. 2009. *Results of archaeological testing at SDI13701 and NAHI-S-1 for the north Agua Hedionda interceptor western segment realignment project, Carlsbad, San Diego County, California.*

Jackson ST, Hobbs RJ. 2009. Ecological restoration in the light of ecological history. *Science* 325:567-568.

Jacobs D, Stein ED, Longcore T. 2010. *Classification of California estuaries based on natural closure patterns: Templates for restoration and management*. Southern California Coastal Water Research Project.

Jacobs RAW. 1986. *Snowy plover (Charadrius alexandrinus). Section 4.4.1, U.S. Army Corps of Engineers wildlife resources management manual*. US Army Corps of Engineers, Portland, OR.

James, GW. 1911. *In and out of the old missions of California: an historical and pictorial account of the Franciscan missions*. Boston: Little, Brown, and Company.

Jennings CW, Strand RG, Rogers TH. 1977. *Geological map of California*. Sacramento, CA: California Division of Mines and Geology.

Jepson WL. 1898. *Erythea: A journal of botany, West American and general*. Berkeley, CA: Dulau & Co.

Johnson JW. 1973. Characteristics and behavior of pacific cost tidal inlets. *Journal of the Waterways Harbors and Coastal Engineering Division* 99(WW3):325-329.

Kelly AO. 1959. *Waters of Agua Hedionda Lagoon reflect the coming of the Spaniards to area*. Carlsbad Journal. *Courtesy of Carlsbad City Library Carlsbad History Room.*

Knox RW. 1933-4. Del Mar, register no. 5410. U.S. Coast and Geodetic Survey (USCGS). *Courtesy of National Oceanic and Atmospheric Administration (NOAA).*

Knox RW. 1934a. Carlsbad, register no. 5412. U.S. Coast and Geodetic Survey (USCGS). *Courtesy of National Oceanic and Atmospheric Administration (NOAA).*

Knox RW. 1934b. *Descriptive report to accompany photo-topographic sheet, register no. T-5412, Ponto to South Oceanside, California, 1934*. U.S. Coast and Geodetic Survey (USCGS). *Courtesy of National Oceanic and Atmospheric Administration (NOAA).*

Knox RW. 1934c. *Descriptive report to accompany topo sheets K, J, I, H, G, B, C, D, E, & F, 1934, Point Loma to Santa Margarita River, California, February to July, 1934*. U.S. Coast and Geodetic Survey (USCGS). *Courtesy of National Oceanic and Atmospheric Administration (NOAA).*

Knox RW. 1934d. Encinitas, register no. 5411. U.S. Coast and Geodetic Survey (USCGS). *Courtesy of National Oceanic and Atmospheric Administration (NOAA).*

Knox RW. 1934-5a. *Descriptive report to accompany photo-topographic sheet, register no. T-5410, Del Mar, California, 1934-35*. U.S. Coast and Geodetic Survey (USCGS). *Courtesy of National Oceanic and Atmospheric Administration (NOAA).*

Knox RW. 1934-5b. *Descriptive report to accompany photo-topographic sheet, register no. T-5411, San Elijo Lagoon to Batiquitos Lagoon, California, 1934-35*. U.S. Coast and Geodetic Survey (USCGS). *Courtesy of National Oceanic and Atmospheric Administration (NOAA).*

Koenen MT, Utych RB, Leslie DMJ. 1996. Methods used to improve least tern and snowy plover nesting success on alkaline flats. *Journal of Field Ornithology* 67(2):281-291.

Kondolf GM, Smeltzer MW, Railsback SF. 2001. Design and performance of a channel reconstruction project in a coastal California gravel-bed stream. *Environmental Management* 28(6):761-776.

Kuhn G, Shepard FP. 1985. Dana Point to the Mexican border. In *Living with the California coast*, ed. Gary Griggs and Lauret Savoy. Durham, NC: Duke University Press.

Kuhn GG, Shepard FP. 1984. *Sea Cliffs, Beaches, and Coastal Valleys of San Diego County: Some Amazing Histories and Some Horrifying Implications*. Berkeley, CA: University of California Press.

Lafferty KD, Swift CC, Ambrose RF. 1999. Extirpation and recolonization in a metapopulation of an endangered fish, the tidewater goby. *Conservation Biology* 13(6):1447-1453.

Lamb JR. 1977. *A brief history of the Batiquitos Lagoon area: local history 43.* Mira Costa College.

Largier JL, Hollibaugh JT, Smith SV. 1997. Seasonally hypersaline estuaries in mediterranean-climate regions. *Estuarine, Coastal, and Shelf Science* 45:789-797.

Laton RW, Foster J, Figuero O, et al. 2002. *Sediment quality and depositional environment of San Elijo Lagoon.* San Elijo Lagoon Conservancy, California State University Fullerton.

Lauriers MRD, Lauriers CG-D. 2006. The Huamalguenos of Isla Cedros, Baja California, as described in Father Miguel Venegas' 1739 manuscript Obras Californias. *Journal of California and Great Basin Anthropology* 26(2):1-30.

Leeds CT. 1946. *Report on small craft harbor possibilities in northern San Diego County.* Los Angeles, CA.

Lichter M, Klein M, Zviely D. 2011. Dynamic morphology of small south-eastern Mediterranean river mouths: a conceptual model. *Earth surface processes and landforms* 36:547-562.

Lightner J. 2013. *San Diego native plants in the 1830s.* San Diego, CA: San Diego Flora.

Los Angeles Herald. 1881. Excursion on the California Southern. November 13. *Courtesy of California Digital Newspaper Collection.*

Los Angeles Herald. 1884. Condition of the California Southern. March 20. *Courtesy of California Digital Newspaper Collection.*

Los Angeles Herald. 1887. The Combination Land Co. October 11. *Courtesy of California Digital Newspaper Collection.*

Los Angeles Herald. 1889. The Pacific slope: Replacing bridges on the California Southern. December 19. *Courtesy of California Digital Newspaper Collection.*

Lowell DL. 1985. The California southern railroad and the growth of San Diego Part 1. *The Journal of San Diego History* 31(4).

Lynch HB. 1931. *Rainfall and stream run-off in Southern California since 1769.* Metropolitan Water District of Southern California, Los Angeles, CA.

Mansfield HB. 1889. Hydrography off coast of California from Sand Ridge to Leucadia. U.S. Coast and Geodetic Survey (USCGS). *Courtesy of National Oceanic and Atmospheric Administration (NOAA).*

Marcus L. 1989. *The coastal wetlands of San Diego County.* Sacramento, CA: California State Coastal Conservancy.

Masters PM, Aiello IW. 2007. Postglacial evolution of coastal environments. In *California prehistory: Colonization, culture, and complexity*, ed. Terry L. Jones and Kathryn A. Klar, 35-52. Plymouth, United Kingdom: AltaMira Press.

Masters PM, Gallegos DR. 1997. Environmental change and coastal adaptations in SanDiego County during the Middle Holocene. In *Archaeology of the California Coast during the Middle Holocene*, ed. J.M. Erlandson and M.A. Glassow: Institute of Archaeology, University of California, Los Angeles.

Merkel & Associates, Inc. 2009. *Batiquitos Lagoon long-term biological monitoring program final report.* San Diego, CA.

Meyer HW. 1980. *Pollen evidence for historic sedimentation rates in Batiquitos Lagoon, San Diego County, CA.* Master's thesis, Paleontology, University of California, Berkeley. Berkeley, CA.

Miller JN. 1966. *The present and the past molluscan faunas and environments of four southern California coastal lagoons.* Master's thesis, University of California, San Diego. San Diego, CA.

Miller RG, Miller LB. 1940. [Record of *Eucyclogobius newberryi* from "Lagoon in Agua Hedionda Creek near Carlsbad"]. University of Michigan Museum of Zoology Fish Collection, Fishnet2 Portal, www.fishnet2.net.

Mudie PJ, Browning B, J. S. 1974. *The natural resources of Los Penasquitos lagoon and recommendations for use and development.* California Department of Fish and Game. Sacramento, CA.

Mudie PJ, Browning BM, Speth JW. 1976. *The natural resources of San Dieguito and Batiquitos lagoons.* California Department of Fish and Game, Coastal wetland series, 12. Sacramento, CA.

Mudie PJ, Byrne R. 1980. Pollen evidence for historic sedimentation rates in California coastal marshes. *Estuarine and Coastal Marine Science* 10:305-316.

Murphy TD. 1915. *On sunset highways: a book of motor rambles in California.* Boston, MA: L.C. Page & Company.

NAIP. 2009. [Natural color aerial photos of San Diego County]. Ground resolution: 1m. National Agriculture Imagery Program (NAIP). U.S. Department of Agriculture (USDA), Washington, D.C.

National Weather Service. 2013. Monthly precipitation for San Diego. http://www.wrh.noaa.gov/sgx/climate/san-pcpn.htm.

Neuenschwander LF, Thorsted THJ, Vogl RJ. 1979. The salt marsh and transitional vegetation of Bahia de San Quintin. *Southern California Academy of Sciences* 78(3):163-182.

NMFS (National Marine Fisheries Service). 2012. *Southern California steelhead recovery plan.* National Oceanic and Atmospheric Administration, National Marine Fisheries Service, Southwest Region, Protected Resources Division. Long Beach, CA.

Nordby CS, Zedler JB. 1991. Responses of fish and macrobenthic assemblages to hydrologic disturbances in Tijuana. *Estuaries* 14(1):80-93.

Nordstrom CE, Inman DL. 1973. Beach and Cliff erosion in San Diego County. In *Studies on the geology and geologic hazards of the greater San Diego area, California*, ed. Arnold Ross and Robert J. Dowlen. San Diego, CA: San Diego Association of Geologists.

Oceanside Blade. 1912. City trustees. April 27. *Courtesy of Oceanside Historical Society.*

Oceanside Blade-Tribune. 1928. Coast lagoons proposed parks. June 26. *Courtesy of Oceanside Public Library.*

Oceanside Blade-Tribune. 1931. Water breaks through both local sloughs: Narrow channels cut for drainage start big rushes to ocean. December 17. *Courtesy of Oceanside Historical Society.*

O'Connell L. 1987. A short history of Encinitas. *Courtesy of Encinitas Historical Society.*

Ohara L. 2011. Agua Hedionda Lagoon ER, Batiquitos Lagoon ER, Buena Vista Creek ER, Buena Vista Lagoon ER, Carlsbad Highlands ER, San Diego County. California Department of Fish and Game, South Coast Region. http://www.dfg.ca.gov/lands/er/region5/docs/BatiquitosBuenaVistaAguaHedCarlsbadER.pdf.

Orme AR, Griggs G, Revell D, et al. 2011. Beaches changes along the southern Califronia coast during the 20th century: A comparison of natural and human forcing factors. *Shore & Beach* 79(4):38-50.

Osgood JO. 1881a. Exhibit "A" map (in two parts) of the location of the California Southern Railroad [version a]. California Southern Railroad. *Courtesy of Caifornia State Railroad Museum.*

Osgood JO. 1881b. Exhibit "A" map (in two parts) of the location of the California Southern Railroad [version b]. California Southern Railroad. *Courtesy of Caifornia State Railroad Museum.*

Osgood JO. 1881c. Map of the location of the California Southern Railroad. California Southern Railroad. *Courtesy of Caifornia State Railroad Museum.*

Pacific Rural Press. 1880. A trip down the coast. December 18. *Courtesy of California Digital Newspaper Collection.*

Pascoe J. 1869. *Field notes of the subdivision lines of Township 11 South Range 4 West, San Bernardino Meridian, California.* U.S. Department of the Interior, Bureau of Land Management Rectangular Survey, California. Book 170-16. *Courtesy of Bureau of Land Management, Sacramento, California.*

Patsch K, Griggs G. 2006. *Littoral cells, sand budgets, and beaches: understanding California's shoreline.* Santa Cruz, CA: Institute of Marine Sciences, UC Santa Cruz.

Peinado M, Alcaraz F, Delgadillo J, et al. 1994. The coastal salt marshes of California and Baja California. *Vegetatio* 110(1):55-66.

Pennings SC, Bertness MD. 1999. Using latitudinal variation to examine effects of climate on coastal salt marsh pattern and process. *Current topics in wetland biogeochemistry* 3:100-111.

Pennings SC, Bertness MD. 2001. Salt marsh communities. In *Marine Community Ecology*, ed., 289-316. Sunderland, MA: Sinauer Associates.

Pennings SC, Callaway RM. 1992. Salt marsh plant zonation: The relative importance of competition and physical factors. *The Ecological Society of America* 73(2):681-690.

Pennings SC, Grant M-B, Bertness MD. 2005. Plant zonation in low-latitude salt marshes: disentangling the roles of flooding, salinity and competition. *Journal of Ecology* 93:159-167.

Phillips RP, Bradshaw JS, Gayman W, et al. 1978. *Tidal aspects of Batiquitos Lagoon 1850 to present.* Environmental Studies Laboratory of the University of San Diego.

Phleger FB, Ewing GC. 1962. Sedimentology and oceanography of coastal lagoons in Baja California, Mexico. *Geological Society of America Bulletin* 73(2):145-182.

Pierre C, Ortlieb L, Person A. 1984. Supratidal evaporitic dolomite at Ojo De Liebre Lagoon: Mineralogical and isotopic arguments for primary crystallization. *Journal of Sedimentary Petrology* 54(4):1049-1061.

Pirazzini AP. 1938. *San Diego County.* State of California, Department of Natural Resources, Division of Forestry.

Poole CH. 1855. [Report from San Diego County Surveyor]. In Annual report of the Surveyor-General of the State of California, S. H. Marlette Surveyor-General. James Allen, State Printer. *Courtesy of California State Lands Commission.*

Pope KO. 2004. The geoarchaeology of the southern flank of San Elijo Lagoon. In *Results of NSF-funded archaeological and paleoenvironmental investigations at San Elijo Lagoon, San Diego County, California*, ed. Brian F. Byrd, Kevin O. Pope, and Seetha N. Reddy. Carlsbad, CA: ASM Affiliates, Inc.

Post WS. 1913. *Report on overflow lands at the mouth of San Dieguito River.* Ed Fletcher Papers, 1870-1955 (MSS 81), Box 40, Folder 15. *Courtesy of Mandeville Special Collections, UC San Diego.*

Prestegaard KL. 1975. *Steam and lagoon channels of the Los Penasquitos Watershed, California, with an evaluation of possible effects of proposed urbanization.* Thesis, Master of Science, Geology, University of California, Berkeley.

Purer EA. 1942. Plant ecology of the coastal salt marshlands of San Diego County, California. *Ecological Monographs* 12(1):81-111.

Purser BH. 1973. *The Persian Gulf: Holocene carbonate sedimentation and diagenesis in a shallow epicontinental sea.* New York, NY: Springer-Verlag.

Rechnitzer AB. 1954. Status of the Wood Ibis in San Diego County, California. *The Condor* 56:309-310.

Rechnitzer AB. 1956. Foraging habits and local movements of the wood ibis in San Diego County, California. *The Condor* 58(6):427-432.

Reineck HE, Singh IB. 1973. Coastal lagoons. In *Depositional sedimentary environments: with reference to terrigenous clastics,* ed., 350-354. New York, NY: Springer-Verlag.

Rhemtulla JM, Mladenoff DJ. 2007. Why history matters in landscape ecology. *Landscape ecology* 22(1-3).

Rich A, Keller EA. 2012. Watershed controls on the geomorphology of small coastal lagoons in an active tectonic environment. *Estuaries and coasts* 35:183-189.

Ritter JR. 1963. *Sedimentation in Agua Hedionda Lagoon.* University of California, Berkeley.

Rodgers AF. 1887-8a. *Descriptive report to accompany original field sheet, entitled: topography, Pacific Coast, northward from San Dieguito Valley, California, 1887-8 [T-1898].* U.S. Coast and Geodetic Survey (USCGS). *Courtesy of National Oceanic and Atmospheric Administration (NOAA).*

Rodgers AF. 1887-8b. *Descriptive report to accompany original field sheet, entitled: topography, Pacific Coast, northward from San Marcos Valley, California, 1887-8 [T-1899].* U.S. Coast and Geodetic Survey (USCGS). *Courtesy of National Oceanic and Atmospheric Administration (NOAA).*

Rodgers AF. 1889. *Descriptive report to accompany original field sheet entitled: topography, Pacific Coast, southward from San Dieguito Valley, California, 1889 [T-2014].* U.S. Coast and Geodetic Survey (USCGS). *Courtesy of National Oceanic and Atmospheric Administration (NOAA).*

Rodgers AF, McGrath JE. 1887-8a. Topography, Pacific coast, northward from San Dieguito Valley, California, register no. 1898. U.S. Coast and Geodetic Survey (USCGS). *Courtesy of National Oceanic and Atmospheric Administration (NOAA).*

Rodgers AF, McGrath JE. 1887-8b. Topography, Pacific coast, northward from San Marcos Valley, register no. 1899. U.S. Coast and Geodetic Survey (USCGS). *Courtesy of National Oceanic and Atmospheric Administration (NOAA).*

Rodgers AF, Nelson J. 1889. Topography, Pacific coast, southward from San Dieguito Valley, register no. 2014. U.S. Coast and Geodetic Survey (USCGS). *Courtesy of National Oceanic and Atmospheric Administration (NOAA).*

Rodney Stokes Co. 1920. Map of San Diego County, California. *Courtesy of UC Berkeley Earth Science and Map Library.*

Roth & Associates. 1990. *Cultural resources survey city of Carlsbad.* Carlsbad, CA.

Rowntree LB. 1985. A crop-based rainfall chronology for pre-instrumental record southern California. *Climate Change* 7:327-341.

Roy PS, Williams RJ, Jones AR, et al. 2001. Stucture and function of south-east Australian estuaries. *Estuarine, Coastal, and Shelf Science* 53:351-384.

Rumsey & King. 1910. Map of Cardiff. San Diego County, CA. *Courtesy of San Diego Cartographic Services.*

Safford H, North M, Meyer M. 2012a. Climate change and the relevance of historical forest conditions. In *Managing Sierra Nevada forests. General Technical Report PSW-GTR-237,* ed. Malcolm North, 23-45. Albany, CA: U.S. Department of Agriculture, Forest Service, Pacific Southwest Research Station.

Safford H, Wiens JA, Hayward GD. 2012b. The growing importance of the past in managing ecosystems of the future. In *Historical environmental variation in conservation and natural resource management,* ed. John Wiens, Gregory Hayward, Hugh Safford, and Catherine Giffen, 319-327. Oxford, UK: John Wiley & Sons.

San Diego County. 1928. Aerial photographs, flown between Nov. 1928 & Mar. 1929. *Courtesy of County of San Diego, Department of Public Works.*

San Diego County Planning Commission. 1947. Proposed small boat harbors, San Diego County coastal region, Del Mar to Oceanside. *Courtesy of San Diego Cartographic Services.*

San Diego Regional Water Quality Control Board. 1967. *Water quality control policy for coastal lagoons in San Diego County and Southeast Orange County. Courtesy of Carlsbad City Library Carlsbad History Room.*

San Diego Water Board (California Regional Water Quality Control Board, San Diego Region). 2011. *Los Peñasquitos Lagoon sedimentation/siltation TMDL*. State Water Resources Control Board, San Diego, CA.

San Elijo Lagoon Conservancy. 2005. *The Escondido Creek Watershed restoration action strategy*. Encinitas, CA.

San Francisco Call. 1901. To make salt from sea water. February 12. *Courtesy of California Digital Newspaper Collection*.

San Francisco Call. 1905. Cool weather follows storm: falling temperatures are reported from southern portion of the state. February 7. *Courtesy of California Digital Newspaper Collection*.

Schwartz & Ewing. ca. 1915. Del Mar. South Coast Land Co. *Courtesy of San Diego History Center*.

Scott DB, Mudie PJ, Bradshaw JS. 2011. Coastal evolution of southern California as interpreted from benthic foraminifera, ostracodes, and pollen. *Journal of Foraminiferal Research* 41(3):285-307.

SDG&E (San Diego Gas & Electric). 2013. San Dieguito wetlands restoration. http://www.sdge.com/environment/preservation-properties/san-dieguito-wetlands-restoration.

Shalowitz AL. 1964. *Shore and sea boundaries, with special reference to the interpretation and use of Coast and Geodetic Survey data, United States*. U.S. Department of Commerce, Coast and Geodetic Survey. [Washington]: Government Printing Office.

Sherman L. 2001. *A history of north San Diego County: from mission to millenium*. Carlsbad, CA: Heritage Media Corporation.

Shipman H. 2008. *A geomorphic classification of Puget Sound nearshore landforms*. Puget Sound Nearshore Partnership Report No. 2008-01. Seattle District, U.S. Army Corps of Engineers, Seattle, WA.

Sipe RJ. 1934. Hydrographic survey No. 5663, Encinitas to Carlsbad, Southern California coast. U. S. Coast and Geodetic Survey. *Courtesy of National Oceanic and Atmospheric Administration (NOAA)*.

Smith WCS. 1849. *Narrative of a 49-er and incidents of travel from New York to San Francisco*. [no publisher given].

Smythe WE. 1908. *History of San Diego, 1542-1908*. San Diego: The History Company.

South Coast Land Co. 1912. *Del Mar, California*. Los Angeles, CA.

South Coast Land Co. 1913. Map of mouth of San Dieguito River near Del Mar, California. Inventory of the Charles H. Lee Papers, bulk 1912-1955, MS 76/1 33n. *Courtesy of Holdings of Special Collections & Archives, UC Riverside*.

Stanbro PW. 1971. *Buena Vista lagoon and its use*. Master's thesis, San Diego State College, San Diego, CA.

Stanley-Brown J. 1896. *Bulletin of the Geological Society of America*. Rochester: The Society.

State Coastal Conservancy. 1987. *Revised draft: Batiquitos Lagoon enhancement plan*. California State Coastal Conservancy.

State Coastal Conservancy and City of Del Mar. 1979. *San Dieguito Lagoon resource enhancement program*.

States Publishing Co., Ltd. 1930. Southern California at a glance: History, romance, maps, facts, statistics. Los Angeles, CA. *Courtesy of San Diego History Center*.

Stein ED, Cayce K, Salomon M, et al. 2014. Wetlands of the southern California coast: historical extent and change over time. Southern California Coastal Water Research Project, SCCWRP Technical Report #826, Costa Mesa, CA.

Stein ED, Dark S, Longcore T, et al. 2010. Historical ecology as a tool for assessing landscape change and informing wetland restoration priorities. *Wetlands* 30:589-601.

Storie RE, Carpenter EJ. 1929a. Soil map, Oceanside area, California. U.S. Government Printing Office. U.S. Bureau of Chemistry and Soils, California Agricultural Experiment Station.

Storie RE, Carpenter EJ. 1929b. *Soil survey of the Oceanside area, California*. Soil survey series, 1929, no. 11. Washington, D.C.: U.S. Government Printing Office. U.S. Bureau of Chemistry and Soils., California Agricultural Experiment Station.

Sullivan G. 2001. Appendix 4. Habitat and elevational distribution of salt marsh plant species. In *Handbook for restoring tidal wetlands*, ed. Zedler JB. Boca Raton, LA: CRC Press.

Swetnam TW, Allen CD, Betancourt JL. 1999. Applied historical ecology: Using the past to manage for the future. *Ecological Applications* 9(4):1189-1206.

Swift CC, Nelson JL, Maslow C, et al. 1989. Biology and distribuation of the tidewater goby, *Eucyclogobius newberryi* (pisces: gobiidae) of California. *Contributions in Science* 404:1-19.

Taft OW, Colwell MA, Isola CR, et al. 2002. Waterbird responses to experimental drawdown: implications for the multispecies management of wetland mosaics. *Journal of Applied Ecology* 39(6):987-1001.

Tenaglia NK. 1999. *History of the Buena Vista Lagoon*. http://buenavistalagoon.org/Files_Main/HistoryBVL_NancyTenaglia.pdf.

Teske PR, Wooldridge T. 2001. A comparison of the macrobenthic faunas of permanently open and temporarily open/closed South African estuaries. *Hydrobiologia* 464:227-243.

Tessler R. 1991. Batiquitos battle. *Los Angeles Times*. November 10. *Courtesy of San Diego Public Library.*

Timbrook J. 2007. *Chumash ethnobotany: plant knowledge among the Chumash people of southern California.* Berkeley, CA: Heyday Books.

Trimble SW. 1995. The cow as a geomorphic agent – a critical review. *Geomorphology* 13:233-253.

Tucker W, Bujkovsky G. 2009. *Cardiff-by-the-Sea.* San Francisco, CA: Arcadia Publishing.

Unknown. 1881. Map showing preliminary surveys along the coast from Rose's Cañon to Santa Margarita. California Southern Railroad. Drawing #C 5 1635, ID #33470. *Courtesy of California State Railroad Museum.*

Unknown. ca. 1881a. Map of the location of the California Southern Railroad through San Diego County. California Southern Railroad. Drawing #B 2 624, ID #33447. *Courtesy of California State Railroad Museum.*

Unknown. ca. 1881b. Part 6 Map of Southern California Railway from Fallbrook National City. *Courtesy of San Diego Cartographic Services.*

Unknown. 1888a. Del Mar cut off line "K", location from Soledad Valley to San Dieguito Valley via Ocean Beach. California Southern Railroad. Drawing #B 2 1342 3, ID #33468. *Courtesy of California State Railroad Museum.*

Unknown. 1888b. Location survey for California Southern Railroad Co. from Soledad Valley to San Dieguito Valley along Ocean Beach. Drawing #B 2 1342 1, ID #33468. *Courtesy of California State Railroad Museum.*

Unknown. 1888c. Location survey for California Southern Railroad Co. from Soledad Valley to San Dieguito Valley Del Mar cut off. Drawing #B 2 1342 2, ID #33468. *Courtesy of California State Railroad Museum.*

Unknown. ca. 1900. Torrey Pines looking north [photo]. Box 300, FEP 1042. *Courtesy of San Diego History Center.*

Unknown. ca. 1907-14. "Torrey Pines," La Jolla, Cal. [postcard]. Los Angeles, CA: Newman Post Card Co. *Courtesy of San Diego History Center.*

Unknown. 1913. Del Mar [photo]. Box 242: 2/5, 91-18564-205. *Courtesy of San Diego History Center.*

Unknown. ca. 1915a. Del Mar -- composite panorama of rivers and streams [photo]. Box 334, 4887 B&C. *Courtesy of San Diego History Center.*

Unknown. ca. 1915b. Del Mar: View from Torrey Pines [photo]. Box 300: 1, 80:6532. *Courtesy of San Diego History Center.*

Unknown. ca. 1925a. Looking north [Los Peñasquitos Lagoon]. Box 300: 2/2, FEP 4339. *Courtesy of San Diego History Center.*

Unknown. ca. 1925b. North County Agriculture [photo]. Agriculture Box 19, 90:18138-665. *Courtesy of San Diego History Center.*

Unknown. 1927. Flood damage to Hwy 101 bridge [San Dieguito River at Del Mar Feb 1927]. Box 332: 2/3, 9740 Floods. *Courtesy of San Diego History Center.*

Unknown. ca. 1932. The Torrey pines, San Diego, Calif [postcard]. Los Angeles, CA: Western Publishing & Novelty Co. *Courtesy of San Diego History Center.*

Unknown. 1939. [Untitled aerial photo, April 16, 1939]. *Courtesy of San Elijo Lagoon Conservancy.*

Unknown. 1955. [Untitled oblique aerial photo looking north at western portion of Batiquitos Lagoon]. Photo #A-17. *Courtesy of San Dieguito Heritage Museum.*

Unknown. 1975. *San Dieguito, an illustrated history.* Bonsall, CA: Little Jack Co.

USACE (U.S. Army Corps of Engineers). 1973. *Flood plain information: Buena Vista Creek, Pacific Ocean to Vista, San Diego County, California.* Los Angeles, CA.

U.S. Census. 1853. *The Seventh Census of the United States: 1850.* Washington, D.C.: Robert Armstrong, Public Printer. 965-985.

U.S. Census. 1864. *Agriculture of The United States in 1860; compiled from the original returns of the Eighth Census, under the direction of the Secretary of the Interior.* Washington, D.C.: Government Printing Office.

U.S. Census. 1872. *Ninth Census -- Volume III. The statistics of the wealth and industry of the United States.* U.S. Census Office, Washington, D.C.: Government Printing Office.

U.S. Census. 1882. *Report on the productions of agriculture as recorded at the Tenth Census (June 1, 1880).* U.S. Census Office, Washington, D.C.: Government Printing Office.

U.S. Census. 1895. *Report on the statistics of agriculture in the United States at the Eleventh Census: 1890.* U.S. Census Office, Washington, D.C.: Government Printing Office.

USDC (U.S. District Court, Southern District). ca. 1840a. [Diseño del Rancho Los Vallecitos de San Marcos: Calif.]. Land Case Map B-1210. *Courtesy of The Bancroft Library, UC Berkeley.*

USDC (U.S. District Court, Southern District). ca. 1840b. Plano del paraje de la Agua Hedionda: [Calif.] Land Case Map D-1273. *Courtesy of The Bancroft Library, UC Berkeley.*

USDC (U.S. District Court, Southern District). ca. 1869. U.S. v. The President and Trustees of the City of San Diego, Land Case No. 390 SD [San Diego Pueblo Lands], docket 589. *Courtesy of The Bancroft Library, UC Berkeley.*

USFWS (U.S. Fish and Wildlife Service). 2005. *Recovery plan for the tidewater goby (Eucyclogobius newberryi).* U.S. Fish and Wildlife Service, Portland, OR.

USFWS (U.S. Fish and Wildlife Service). 2007. *Five year review: Tidewater goby (Eucyclogobius newberryi).* U.S. Fish and Wildlife Service, Ventura, CA.

USGS (U. S. Geological Survey). 1891. Oceanside Quadrangle, California: 15-minute series (Topographic).

USGS (U. S. Geological Survey). 1893. Oceanside Quadrangle, California: 15-minute series (Topographic).

USGS (U. S. Geological Survey). [1891]1898. Oceanside Quadrangle, California: 15-minute series (Topographic).

USGS (U. S. Geological Survey). 1901a. Escondido Quadrangle, California: 15-minute series (Topographic).

USGS (U. S. Geological Survey). 1901b. San Luis Rey Quadrangle, California: 30-minute series (Topographic).

USGS (U. S. Geological Survey). 1903. La Jolla Quadrangle, California: 15-minute series (Topographic).

USGS (U. S. Geological Survey). 1930. La Jolla Quadrangle, California: 15-minute series (Topographic).

U.S. Surveyor General's Office. 1875. Map of Fractional Township No. 12 South, Range No. 4 West (San Bernardino Meridian). San Francisco, CA. *Courtesy of Bureau of Land Management.*

U.S. Surveyor General's Office. 1876. Map of Fractional Township No. 14 South, Range No. 4 West (San Bernardino Meridian). San Francisco, CA. *Courtesy of Bureau of Land Management.*

U.S. Surveyor General's Office. 1881. Map of Township No. 13 South, Range No. 4 West (San Bernardino Meridian). San Francisco, CA. *Courtesy of Bureau of Land Management.*

U.S. Surveyor General's Office. 1883. Map of Fractional Township No. 12 South, Range No. 4 West (San Bernardino Meridian). San Francisco, CA. *Courtesy of Bureau of Land Management.*

U.S. Surveyor General's Office. 1890. Map of Fractional Township No. 12 South, Range No. 4 West (San Bernardino Meridian). San Francisco, CA. *Courtesy of Bureau of Land Management.*

Van de Hoek RJ. 2006. *Edith Abigail Purer.* http://www.naturespeace.org/pureranthology.htm.

Van Dyke TS. 1887. *County of San Diego: The Italy of southern California.* National City, CA: National City Record Steam Print.

Van Dyke TS, Leberthon TT, Taylor A. 1888. *The city and county of San Diego: illustrated and containing biographical sketches of prominent men and pioneers.* San Diego, CA Leberthon & Taylor.

Ver Planck WE. 1958. *Salt in California.* State of California, Department of Natural Resources, Division of Mines. San Francisco, CA.

Vogl RJ. 1966. Salt-marsh vegetation of Upper Newport Bay, California. *Ecology* 47(1):80-87.

W.W. Elliott & Co. 1883. *History of San Bernardino County, California including biographical sketches.* San Francisco, CA: W.W. Elliott & Co.

Wade SA, Wormer SRV, Thomas H. 2009. *240 years of ranching: Historical research, field surveys, oral interviews, significance criteria, and management recommendations for ranching districts and sites in the San Diego region.* California State Parks.

Waisanen PJ, Bliss NB. 2002. Changes in population and agricultural land in conterminous United States counties, 1790 to 1997. *Global Biogeochemical Cycles* 16(4):1-18.

Warme JE. 1971. *Paleoecological aspects of a modern coastal lagoon.* University of California Publications in Geological Sciences, 87. Berkeley, CA: University of California Press.

Warren CN, Pavesic MG. 1963. *Shell midden analysis of site SDi-603 and ecological implications for cultural development of Batiquitos Lagoon, San Diego County, California.* Archaeological Survey Annual Report 1962-3: 323-406. University of California, Los Angeles.

Webb CK, Stow DA, Chang HH. 1991. Morphodynamics of southern California inlets. *Journal of Coastal Research* 7(1):167-187.

Weis DA, Callaway JC, Gersberg RM. 2001. Vertical accretion rates and heavy metal chronologies in wetland sediments of the Tijuana Estuary. *Estuaries* 24(6A):840-850.

Welker ST, Patton RT. 1995. *San Elijo Lagoon Ecological Reserve biological element for master plan/management plan.* County of San Diego, Parks and Recreation Department. San Diego, CA.

West JM. 2001. San Dieguito Lagoon. In *Handbook for restoring tidal wetlands*, ed. Joy B. Zedler, 22-23. Boca Raton, LA: CRC Press.

Western Regional Climate Center. 2013. San Diego WSO Airport, California: Monthly total precipitation. http://www.wrcc.dri.edu/cgi-bin/cliMONtpre.pl?ca7740.

Wheeler MC, Copeland F, Lockling LL. 1872. Official map of the western portion of San Diego County, California. San Diego County. *Courtesy of San Diego Cartographic Services.*

Wheeler MG. 1874-5. *Field notes of the subdivision lines of fractional Township 12 South, Range 4 West, San Bernardino Meridian.* U.S. Department of the Interior, Bureau of Land Management Rectangular Survey, California. Book 170-17. *Courtesy of Bureau of Land Management, Sacramento, California.*

White MD, Greer KA. 2006. The effects of watershed urbanization on the stream hydrology and riparian vegetation of Los Peñasquitos Creek, California. *Landscape and Urban Planning* 74(2):125-138.

Whitfield AK. 1992. A characterization of southern African estuarine systems. *Southern African Journal of Aquatic Sciences* 18(1/2):89-103.

Williams GD. 1996. *The physical, chemical, and biological monitoring of Los Peñasquitos Lagoon.* San Diego State University, Pacific Estuarine Research Laboratory (PERL). San Diego, CA. http://www.perl.sdsu.edu/Reports/LPL95-96.pdf.

Williams GD, Zedler JB. 1999. Fish assemblage composition in constructed and natural tidal marshes of San Diego Bay: relative influence of channel morphology and restoration history. *Estuaries* 22(3):702-716.

Williams PB, Orr MK, Garrity NJ. 2002. Hydraulic geometry: a geomorphic design tool for tidal marsh channel evolution in wetland restoration projects. *Restoration Ecology* 10:577-590.

Williamson RS. 1861. General map of explorations and surveys in California. Washington, D.C.: T. Ford XII. *Courtesy of David Rumsey Map Collection.*

Wilson W. 1883. *History of San Diego County, California, with illustrations, descriptive of its scenery, farms, residences, public buildings... from original drawings, with biographical sketches.* San Francisco, CA: W.W. Elliott & Co.

Wood BD. 1913. *Gazetteer of surface waters.* Department of the Interior, United States Geological Survey. Washington, D.C.: Government Printing Office. 243.

WyGISC (Wyoming Geographic Information Science Center). 2008. California Watershed Boundary Dataset (WBD).

Yechieli Y, Wood WW. 2002. Hydrogeologic processes in saline systems: Playas, sabkhas, and saline lakes. *Earth-Science Reviews* 58:343-365.

Young AP, Ashford SA. 2006. Application of airborne LIDAR for seacliff volumetric change and beach-sediment budget contributions. *Journal of Coastal Research* 22(2):307-318.

Zedler JB. 2001. *Handbook for restoring tidal wetlands.* Boca Raton, LA: CRC Press.

Zedler JB. 2012. Diverse perspectives on tidal marshes. In *Ecology, conservation, and restoratin of tidal marshes*, ed. Arnas Palaima, 265. Los Angeles, CA: University of California Press.

Zedler JB, Callaway JC, Desmond JS, et al. 1999. Californian salt-marsh vegetation: An improved model of spatial pattern. *Ecosystems* 2:19-35.

Zepeda-Herman C. 2009. *Archaeological monitoring and feature excavation for the San Elijo Lagoon Nature Center Project, San Diego County, California.* RECON Environmental, Inc., San Diego, CA.

www.ingramcontent.com/pod-product-compliance
Lightning Source LLC
Chambersburg PA
CBHW041727210326

41598CB00008B/798